*Asian Psychology*

# ASIAN

*Edited by* GARDNER MURPHY

*and* LOIS B. MURPHY

Basic Books, Inc., Publishers

# PSYCHOLOGY

NEW YORK   LONDON

TO THE CHILDREN OF

The Sarabhais,
The Kothurkars,
The Charis,

The Hsus,
The Chens,

The Tsushimas,

And the Asian Psychology of Tomorrow

Library of Congress Catalog Card Number: 68–22856
Manufactured in the United States of America
Book designed by Jacqueline Schuman
SBN 465–09504–6
72737475   10987654321

# Editors' Introduction

This is the first of a series of volumes which we hope to edit in the next few years dealing with: Asian Psychology, Western Psychology from the Greeks to William James, Developmental and Psychoanalytic Psychologies, Altered Ego States, and one or two others. These books are planned as a series in terms of articulation of the various volumes with one another and in the hope of obtaining reasonably good coverage without excessive overlapping. The aim, in fact, is to see, in as broad a context as possible, the human effort to understand the mind—or, if you like, mind, heart, will, the person as a whole—in systematic terms.

There is, in a sense, a universal psychology—a psychology accepted by all men and used in dealing with ultimate things as well as with the things of the marketplace. There are what Fritz Heider calls "common sense psychologies," implicit, taken-for-granted psychologies. There is, for example, the pleasure principle: no one would do anything contrary to his own best interests. There is always a lurking motive, a goal, or a woman to be won, punishment or shame to be avoided. Another "common sense" notion is that things come to be associated in the mind if they have personal significance. The hunting horn and the dinner bell, for example, take on melody more than is their natural share, and the form and features of those who have injured us become gradually more hideous.

There is likewise in the common or universal psychology a psychology of man and of woman, a psychology indeed of old man and woman, young man and woman, a psychology of childhood and adolescence, a psychology of the proud and of the humble, or, as Nietzsche put it, "master morality and slave morality." There is, in other words, the beginning everywhere of a system of psychology, a system of assumptions which, because we are human, must take essentially similar human form all over the face of the globe. Man takes this so much for granted that his folktales, his plays, and his verse repeat the theme everywhere, and when he makes up for his children or for his own fireside hours stories to delight the soul, there are animal characters—bears, lions, jackdaws, tortoises, elephants, dogs, and grasshoppers—who, in their wisdom or foolishness, nobility or baseness, accept human attributes, talk as men do, and convey to us, through a slightly altered guise which reflects their own animal personalities, the basic wisdoms and cynicisms of humankind. When man draws these animals close to himself, he gives human names and attributes to them exactly as he does to trees and stars and creates the world of "animism" in which animal attributes are applied to nonliving things and, in the last analysis, human attributes are applied to both.

There is, then, a sort of universal psychology, a conception of mind and character, with its broad and universal attributes and its capacity for individualization in each case, which would permit the writing of the *universal psychology*—a psychology that applies wherever mind, or let us say, in intuitively accepted terms, wherever mind, heart, and will exist. And, as we have noted, Fritz Heider and indeed others now emulating him have striven to find the common-sense assumptions about psychology, the things taken for granted in all human living. We are maintaining that there really is a universal psychology of this sort; indeed, in two senses such a psychology has relevance for us now as we start on this big task of ours. There is a universal psychological system, and we shall attempt to find it appearing and reappearing like a form in the shadows or a rainbow in the glistening droplets after a storm. Or, perhaps better, we shall find it like the sun which the ancients imagined to move from one "house" to another during the course of a year, the zodiac representing the "houses of the sun"; in this sense there would be one sun but many appearances—spring, summer, autumn, and winter as the sun strengthens, weakens, and changes hue and form depending upon its place in the structure of the world.

But there is another sense in which there is a universal psychology. There are not only assumptions; there are observations, facts with relative objectivity, contours of reality which a gradually increasing awareness of *method* may help people to grasp and apply.

Though the conception of a scientific psychology has been with us only about a century, there has been, in the East and in the West, in ancient and in modern times, an almost endless variety of efforts to become *systematic;* to learn to observe, to order, to integrate, and to interpret—to make, in other words, a science out of the things of the mind. In every society there are wise men regarded as "past masters in the things of the spirit" and in some societies wisdom and prophecy are gradually replaced, as they have been in physics and biology, by system, by order, and by a conception as to how system and order can really be found.

There is, then, a universal psychology both in the sense that there really are uniformities over the face of the globe and in the sense that there are uniformities in the *ways of studying and interpreting these psychological realities* everywhere displayed. Clearly it will be our task in these volumes to help the reader to see how much uniformity there is, what the deep resemblances are between the universal principles discovered in India, in China, in Japan, in Greece, in medieval Europe. We must look for uniformities in method of approach, whether the methods of stargazer or mathematician, physician or laboratory scientist, and at the same time be forever aware of cultural diversities, historical uniqueness, the special insight or the special gift of observation which characterizes one historical people as distinctive and different from all others. We shall be looking for the universal psychology, but we shall also be looking for psychologies, systems of thought about thought, so to speak, local and specific insights.

In the college courses of fifty years ago—and, in fact, here and there in some still surviving course catalogues—one encounters psychology as a subdepartment within the department of philosophy.

Or, indeed, one encounters a department of philosophy and psychology without subordination of one to the other. This has occasioned much thought among the young who have battled their way through the problems of the curriculum builders; who learned how the languages were grouped together as Romance, or Germanic, how the sciences were classified as physical sciences, biological sciences, and social sciences. Was psychology a biological science or a social science, or both, or neither? Watching closed-circuit television today, they see laboratory demonstrations of studies of emotion and learning in mice and rats, or even in monkeys and apes, regard psychology as a biological science, and notice here and there evidence that it has a different status in the minds of some individuals, some colleges, some entire systems of educational theory.

This is intelligible in historical terms. It turned out, on inspection, that there were three quite different ways of viewing this intimate relation of philosophy and psychology. First, the commonest way of explaining the matter was that philosophy was the main great trunk of learning and that from this branched off the physical sciences, so that there was a natural philosophy, or philosophy of nature; farther on up the trunk there branched off the biological sciences, the "natural history" of our grandfather's time, the study of plants and animals; and then much farther up the great trunk there appeared subdivisions of philosophy which dealt with the mind and its ways and the ways of the ethical principle in man, giving rise to what was offered in American universities as "mental and moral philosophy" or, for short, "mental and moral." In the German-speaking world the "sciences of the spirit," or the "moral and social sciences," had already split off from the "natural sciences," and the tradition in the English-speaking world reflected this Germanic development although it had its own independent supporting background as well.

First, then, psychology was a part of philosophy because it was immature; it had not yet completely broken away from the great parent trunk. It was eminently reasonable from this viewpoint that one should study Plato and Aristotle in order to understand such great modern figures of philosophy as Locke, Leibnitz, and Kant and important that one understand Locke, Leibnitz, and Kant to be able to navigate in nineteenth-century thought. One then had a modest place for the new experimental psychology which was arising side by side with the evolutionary biology of Darwin, which aspired to be a "natural science"; within its enfolding arms a laboratory was, from 1879 onward (the date of Wilhelm Wundt's laboratory of experimental psychology), available as a signal of the new scientific era.

But there was a second reason for the relations of philosophy and psychology, namely that the philosopher took universal knowledge and wisdom as his goal, and he certainly had to include wisdom about the mind. He undertook, of course, to understand physics and astronomy, biology and medicine; the time came for him to study the ways of the mind also. Herbert Spencer, having conceived a universal plan of human knowledge, treated the physical, then the biological, then the psychological, and finally the social sciences. The philosopher must, of course, give long and earnest attention to the ways of the mind because it is an especially precious object of knowledge.

But there remains a third reason for the intimate relations of the two. Psychology may bend back on itself and study the psychology of the psychologist, and, if it may do this, it may certainly also study the psychology of the philosopher. In other words, attempting to pass beyond the realm of *object* under investigation, it may become the *means* of investigation. The psychologist may be he who understands philosophy (as far as anyone can), because it is through psychological principles that the ways of philosophers can be grasped. As a matter of fact, Hertzberg wrote a vigorous book a few decades ago on just this problem. *The Psychology of Philosophers* is an attempt to show why Thomas Hobbes, who wrote "Myself and fear were born twins," makes society depend first on the fear motive; why William James, reared in the glowing optimism of the evolutionary period, but wracked by nervous suffering, had to create a "pragmatism" which was risky, chancy, practical, making the best of glowing moments in the midst of an inscrutable world whose ways are "just so much *weather*."

Perhaps these paragraphs will explain why we feel no obligation to apologize for the rather considerable amount of attention given in these volumes to the philosophy of the mind and, so to speak, the ways of the mind as one uses the mind in philosophical endeavors. This is a part of a world psychology, and in due proportion it must receive its just consideration.

Perhaps most distinctive about the present volumes is their aspiration to speak a universal language, a common human language, a language in which our common humanity is taken for granted, in which the relative uniformity of our thoughts, feelings, and impulses is seen in terms of some broad conception of the development of life upon the earth, and which provides enough uniformity amid the diversity of human culture to permit some basic unity in the ways in which people think about the mind.

But from the point of view of philosophy there is a paradox here. The fact that there is some uniformity in the ways in which people think, feel, and act does not imply that the explanations they give for thoughts, feelings, and actions are correct; it does not imply that the common or universal psychology of the marketplace is a correct psychology; it does not imply that because ancient Indian or Greek philosophers, or contemporary American or Soviet psychologists, reach identical formulations of the nature of thought or the will, these must necessarily be scientifically sound conceptions. There is a psychology of self-deception which is as rich and complex as the psychology of direct apprehension of external reality. There is the world of illusion, delusion, inner self-deception, seeing things the way we wish to see them, or in panic as we fear they might turn out, a psychology written in terms of what the psychoanalysts call "mechanisms of defense," with which we shall have to come to terms. We could, if we wanted to, go the way of some philosophers —ancient and modern, Eastern and Western—who say that all perception is delusion and that there is no knowable reality. We might go the way of some modern philosophers and say that there is no reality except that which we directly apprehend, control, measure, and use, and that we might as well stop making precious fantasies out of the hard, tough, real road in front of us. Or we might, as lovers of scientific method, try to understand the very process of knowing in terms of the psychology of the knowing process. We might keep ourselves free of both the doctrinaire nihilism, the know-nothing insistence of the one and the confident "positivism" of the other.

Our ultimate purpose as psychologists, however, is not to try to settle for our readers the most difficult of philosophical questions, but rather to show how and why doubts, difficulties, affirmations, and self-deceptions arise and what kind of knowing about ourselves as human beings can be understood in terms of psychological principles themselves.

In attempting a view of worldwide psychology, one must, in the last analysis, classify in terms of specific groups of human beings who have developed specific ways of looking at the mind. There is, for example, ancient India, and there is the contemporary Soviet Union. Ideas are not developed in a cultural vacuum. It takes many men over many generations to develop a mature way of looking at the mind, and it takes a very considerable amount of time and skill to create a good historical and philosophical perspective to represent a system of psychology. Partly because of geographical factors (such as the Himalayan mountain chain between India and China), partly because of cultural-economic-political-military barriers (such as those separating the Christian from the Moslem world in the Middle Ages), systems of ideas have developed within definite cultural areas; these must be respected and studied in their own right. Ancient systems give rise to medieval systems through complex sequences of economic-political-military-philosophical struggles, and medieval systems give rise to modern systems. One's materials come to one in this fashion. If you wander through a museum in which there are tapestries or armor, jade, porcelain, or glass, you find that it is only by well-informed knowledge of the history of these crafts and the ways in which they are borrowed back and forth from one cultural area to another that one can develop a classification, a "taxonomy" to group the things of art as a naturalist groups the things of nature. The psychologist has no choice in such matters. He must take systems of ideas where they have developed.

But does this mean that he does not classify according to the intrinsic or inner message of the psychology itself? For example, suppose the theory of the emotions among some ancient Hindu sages were practically identical with theories developed in the

Western world in the seventeenth century. Would this mean, then, that we would still resolutely group together even the most diverse Indian and seventeenth-century Western ideas simply because they belong to such historical periods? Would it not be more effective to develop a sort of scheme of organization of ideas about psychology, comb them, order them, arrange them according to a sort of museum in which intimate psychological oneness is given priority over all other considerations?

This kind of stubborn dilemma, whether to arrange his material by cultural epochs or by the specific issues dealt with, yields poorly to rational decision. How much cross-classification, how much comparing of thirteenth-century Byzantium with contemporary Tokyo can he carry out without appearing to be trivial or perverse in looking for forced analogies? We do not believe that any ultimately acceptable answer can be found. It is necessary to maintain the integrity of a cultural whole as one talks about China, Japan, Greece, Rome, or the Britain of the eighteenth century. At the same time, it is obviously necessary to weave back and forth between the different systems, pointing out a resemblance here, a contrast there, resorting as much as one may to the use of footnotes, forward and backward references, adding interpretive, interstitial, and summary material to show the connections between the psychologies of different times and places. Ultimately this becomes a matter of taste, for one could invite sheer chaos by weaving too many threads and ribbons back and forth between dozens of psychological systems. We shall do as much as we can. We shall remain, however, primarily concerned to make the integrity of each whole psychological system evident to the reader.

We begin with Asian psychologies largely because they are relatively independent of our own Western tradition, yet essential if we are to appreciate differences. They derive, in one way or another, from some of the insights and some of the psychological and physiological disciplines of preliterate peoples, especially those who sought a special kind of awareness of their own relation to the universe. In India, nearly three thousand years ago, sages and adepts began to develop particular ways of thinking which were expressed in the Upanishads, in yoga, in the Bhagavad-Gita, later in other Hindu systems, and in the revolutionary ideas of Buddhism. Buddhism was carried across the great mountains to China, and ultimately to Japan, taking on new forms as it moved, while, in the meantime, it was spreading similarly to Burma, to Ceylon, to Thailand, and to other parts of the southern and southeastern portions of the great Asian land mass.

Why, not being professional scholars in this field, do we venture upon a task which, in scholarly terms, we know we cannot possibly fulfill? Throughout our professional lives this issue has been with us: Why do we do things which others could more competently do? As before, the reason lies in the fact that others have not done the things we would like to see done and that we hope to suggest new implications; and thus we may build work so that others may erect a better structure. We are well aware that a profound grasp of these Asian systems is not for us, nor for our readers. Indeed it is only for a few who can devote themselves to unrelenting study of these specific issues.

In particular, we want to emphasize that it is by firsthand experience in living according to these systems that one can be said to have mastered them. He who writes about the discipline of yoga, but knows it just in terms of words to be noted, remembered, and discussed, has not really entered into yoga at all. He who follows any of the Buddhist training procedures, whether of another era or of the Zen of recent centuries, knows only in what direction the land lies; he has not seen its horizons or tasted its fruits. Our present efforts may help those who would like to develop a familiarity with the great maps on which distant lands are delineated. It will be the *traveler* who can hope to know these lands.

This is not a survey, a digest, or an interpretation of Asian psychology. It is highly selective. It aims to offer some of the enduring and significant aspects of Asian thought about mind and life, but arbitrary decisions as to what to select are with one constantly in any such task and there are vast areas that simply have to be omitted altogether. Thus it seemed essential to use the classics of Indian psychology; yet the practical psychology of the lawmakers, the moralists, the untutored village philosophers, the wise old women who appear in the rich folklore of India, are hardly represented at all. The wide expanse of Chinese Buddhism had to be compressed within a brief circle, and the exquisite subtleties of Japanese folklore and drama could not be detailed.

Indeed it is mainly efforts at systematic psychology as such that are included here. Though the psychologies of Asia are almost without exception religious psychologies, they are represented only when their task is not mainly religious but psychological as such. To justify the selections would lead to endless quibbling and be as unsatisfactory to us, as editors, as it could ever be to the reader. Let it stand then: These are our selections, with interpretations here and there where we felt they would be helpful.

Even so, there is a further explanation to offer. Why no psychology of the ancient Middle East? Why, for example, no psychology from Zoroaster, and why indeed no psychology from Hebrew psalmists, sages, and prophets? Why nothing from Tibet? Why nothing from the great Islamic tradition which began in the seventh century? The answers lie in our struggle with the problem of classification. Zoroastrianism is a tremendous religious idea, but we do not find in it a psychology. The same is true of the earliest Hebrew records. As they developed, they interacted with other ideas and were less and less Asian. Later, in a volume dealing, in part, with the psychology of the Greeks, we hope to mention the interaction of Hebrew and Greek ideas in the "Hellenistic" period, showing the transition from early to later Hebrew formulations of the nature of man and his place in the world, and fusing it with the early Christian tradition which stems from both Hebrew and Greek sources.

The same conception about the international or intercultural nature of Islamic thought appears when it becomes evident to the reader (as it will in a later study) that the real fulfillment of Islamic psychological ideas comes about when they interact with the ideas of Aristotle. They do not, therefore, belong to Asia in any strict or possessive sense. They represent the region of intersection of certain Asian with certain European ideas.

Moreover, the *modern* psychology of India, China, and Japan, so rich with reference to clinical and experimental concepts, as expressed in the psychological journals published in these countries, is not represented here as "Asian." It is only the indigenous or traditional psychologies of three great cultural traditions of Asia—Indian, Chinese, and Japanese—with which this book is concerned.

*Gardner Murphy*
*Lois B. Murphy*

Topeka, Kansas
April 1968

# Contents

*Asian Psychology*

# Prologue / THE ASIAN SETTING

The first great psychological system, or a system of systems, is that of India, and we shall accordingly let the main show begin in the vast theater of Indian psychology. The task is much too big for any contemporary scholar, though he be an expert in Sanskrit and a historian, linguist, and philosopher of the first order. We have used as consultant our friend Professor B. Kuppuswamy, long of Mysore University, now of the India International Centre in New Delhi, relying heavily upon the translations of Professors S. Radhakrishnan and J. H. Woods, and the excellent modern commentary of C. E. Eliot and of Heinrich Zimmer and Joseph Campbell.

In introducing Indian psychology we shall offer a sketch and a viewpoint regarding the whole historical method. Does history "repeat itself"? Historians are still unable to agree on the universality, the lawfulness, of historical process, on the question whether "history is a science." There are certainly many local uniformities, developments which lead, for example, into certain types of revolutions which have appeared in many scattered points in time and place. Sometimes quite bold and intriguing philosophical conceptions have arisen, such as Arnold Toynbee's conception of "challenge and response," according to which similar challenges in similar situations lead to similar responses and similar historical outcomes over the face of the globe. If such a view of history is sound, we should do all we can to draw a picture of the development of psychology according to an orderly scientific historical principle. We should be able to show that similar psychologies have arisen under similar cultural, socio-economic, and political conditions. Actually, we shall make a few bold attempts at this here and there. On the whole, however, we have not found that vast historical systems, such as that of India, lend themselves well to this method. What we have found, and this is quite a different thing, is that well-known general principles regarding geography, climate, availability of water, protection by mountain chains against both weather and human enemies, availability of trade routes by land and sea, similarity and diversity among neighboring peoples, the possibility of exchange of products, and many other factors need to be applied to the specific, concrete individual case which we know as the historical event. In the case of India, for example, we shall try to show not that Indian psychology had to arise at a particular time and place because of some universal principle, but rather that we can partly understand it by noting a few broad principles relating to the background, the special conditions, which made Indian psychology possible. We attempt, therefore, to sketch such a background.

From the Persian Plateau more than three thousand years ago, waves of migration flowed across the western Himalayas down into the valleys of India, the dark-skinned natives being gradually overwhelmed by the lighter-skinned newcomers who spoke an "Indo-European" or "Aryan" language, akin to the language of Greece and Rome. Culturally, the invaders were a relatively simple people. Their religion seems to have been strong, clear cut, anthropomorphic. Their pantheon had its nature gods, gods of sky and sun, fire and forest. They had not developed a systematic psychology any more than had their cousins the Greeks. They were, however, adept in the ways of war, and they asserted and maintained their military superiority as they slowly moved south and southeast into

India. They early took a position against intermarriage, and, although a great deal of mixture went on, the system of "caste" which developed was a social expression of the desire of the militarily strong and superior to maintain themselves apart from those whom they conquered. Gradually a series of major subdivisions—recognizing the priestly caste, the military caste, and the farm- or land-owning caste, with more and more complicated and subordinate positions imposed on the conquered groups—was established into a great caste system.

It used to be said, and until twenty or thirty years ago could be said with some plausibility, that the dominant or "superior" Aryan peoples gradually developed a new way of life, a new philosophy and psychology; as they settled down on the plains and in the valleys of northwestern India, they lost the buoyant and confident philosophy, the Nietzschean "master morality." They became thoughtful, somber, and serious, and developed the rich contemplative psychology of India which became standardized in the first millennium before Christ, with new and vital derivatives, on until nearly A.D. 1000. From this point of view the dark-skinned non-Aryan peoples had relatively little to contribute; it was the climate, weather, perhaps the unchallenging character of the face of the earth in India, which led the Aryans to a relative loss of exuberance and the development of a relatively withdrawn or, indeed, pessimistic outlook.

But the whole course of archaeological, linguistic, and historical research has moved relentlessly against this point of view. It is perfectly plain today from excavations (such as those at Mohenjo-Daro) that long before the Aryan invasion there was a vast and magnificent culture, developed as early as 2500 B.C., in what is now the region of Pakistan, and indeed probably over a considerable area in western India and in the western part of the Mesopotamian region. It is likewise plain that the conquering peoples got a great deal of their civilization from those whom they conquered; it is likely, therefore, that much of the serious, contemplative, withdrawn, or pessimistic psychology was there in the civilization of India before the Aryan peoples ever arrived. We shall not attempt here the intricate philosophical and historical task of evaluating the

5

pre-Aryan psychology. We doubt whether anyone could do this; certainly we cannot. We can, however, say that the interaction of a buoyant and confident people with a profoundly civilized and less confident cultural group resulted before very long in the development of a psychology, a religious psychology, utterly different from that which is expressed in the earlier period. The great buoyant hymns of the Rig-Veda were replaced by the contemplative, psychologically rich doctrines of the Upanishads (pages 45–72 below). Within a few centuries there developed in the Sanskrit language of the invading Aryan peoples a conception of the mind, or soul, or spirit, rooted in the utterly changeless reality of the inner self, or atman, which may learn of its own existence, which may protect itself against illusion and the passions, which may become the core of a reality which reaches out until ultimately the individual atman and the cosmic spirit are one. From this philosophy of the atman develops a conception of purity, changelessness, nobility, freedom from deception, freedom from passion, deceit, and delusion, which reappears in almost all the forms of Indian philosophy.

At the same time, partly in union with this and partly in contrast with it, there is a practical psychology—a psychology of the marketplace and the court—which comes down to earth and guides the merchant, the lawyer, and the prince, much as the court jester or, at a higher level, Machiavelli guided the practical wielder of power in the Western world.

Indian psychology became in time interwoven with all of the complex life of the Indian subcontinent. No matter how primitive the life of an Indian village may be, the Indians are a psychological people. There is the great tradition of the Upanishads to be encountered everywhere. The philosophical poem of the Gita (pages 76–96) is embedded in one of the great national epics; the little roadside temple and the grand religious ceremonies alike are saturated in the Sanskrit tradition, which indeed has made itself felt all the way down to the tip of India at Cape Comorin. Even though the vernacular language everywhere in southern India is unrelated to the Sanskrit tradition of the north, nevertheless the religion and therefore the psychology of the Sanskrit world continue in the curious form we have just described, which has defined the Indian conception of mind wherever serious thinking about the mind is done.

We shall, then, with the help of Professor Kuppuswamy, survey a number of the psychological systems of India, showing their common roots in the Sanskrit tradition and their distinctive character in contrast to the psychology of any other people.

Awareness of suffering in all its forms is one of the great cardinal attributes of Indian psychology in contrast to the buoyant optimism of the Rig-Veda. The Brahman priests, lineal descendants of the priestly caste who came with the armies into India, have stylized more and more a verbal and ritual expression of the ways by which the human spirit can free itself of desire, of the frustration which comes from defeated desire, of the self-imposed damage which comes from passion and deceit, and have pointed to the only course open to a wise man—namely, to escape from enslavement to the body and cultivate the purity which the atman in its own existence represents. It is plain, however, that this doctrine did not fully satisfy, that it left awareness of suffering as a major feature in the life of India; that poverty, disease, and death caused unassuaged misery; and that there was a place for a prophet who could offer more radical salvation. He appeared in the form of a prince, Lord Gautama, the enlightened one, or the Buddha as we may say, who realized with a special poignancy the misery inherent in life itself, taught a more practical system of principles, the "noble eightfold path" by which common people could hope to transcend the principle of suffering and find, if not in this life then through a series of incarnations, an ultimate absorption into a selfless impersonal state: nirvana.

From the Western point of view Buddhism may well appear to be essentially a development or an exaggeration of the earlier Indian philosophy which we call Hinduism. It is plain, however, that from an Indian point of view Buddhism was both more personal, more radical, and more practical than Hinduism and that it was therefore in competition with it; in fact, the competition continued century after century, so that as Buddhism made its way into Ceylon and Burma, across the Himalayas into China and ultimately to Japan, and to Thailand and other parts of southeastern Asia, it came to be regarded everywhere as completely distinctive. It is the figure of the Buddha and the cult of the Buddha's wisdom that dominates. It is the noble eightfold path rather than the preoccupation with the atman which is distinctive.

We suggested that the geography of Asia had much to do with Asian psychology. But if we wish we may push this geographical interpretation much further, with special reference to Buddha's compassion in suffering. The tropics, as one scans them through Africa, South America, Indonesia, and Southeast Asia, contain many areas of enormous poverty, enormous distress, and often, historically speaking, enormous despair. The man from temperate climes is often overwhelmed by the sheer mass of despondent and seemingly helpless humanity. Religions of escape, of other-worldliness, of magnification of eternal good as against the physically inescapable evil, have often led to the belittlement of individuality except insofar as the prophet may offer, like Lord Buddha, a collective escape through cutting the cords which bind us to this physical here and now, so fraught with frustrated desires, poverty, disease, and death.

Moreover, the fact of loneliness in the midst of the throng of fellow sufferers is a familiar theme. David Riesman's conception of "the lonely crowd" is the most recent and, to us, the most vivid delineation of a kind of emptiness which supervenes when hope is gone; in sharp contrast to the sense of fellowship and group life and power which maintains the struggle when once there is *hope* (see page 112). The common hopelessness of humanity in its suffering polarizes one toward the prophet and toward the eternal peace and joy which are found when once escape from the engulfing common predicament is a real possibility.

There is no way of "proving" the justice of this thesis. It has, of course, often been proposed in one form or another. We bring it up here because it is seldom clearly stated as an alternative to the usual views as to why the post-Vedic religions of India are so other-worldly, so somber with respect to this world, and so confident regarding another kind of existence.

It would be a very nice question to try to define whether the belief in karma (page 38) and the cycle of rebirth is a sort of "cause" or a sort of "effect" of the type of pessimism which we have tried to describe. One might argue that the man caught in a hopeless web of miseries can always imagine, simply because "hope springs eternal in the human breast," a better world after rebirth, ignoring the considerable risk that the next incarnation will be worse rather than better. One can, however, also argue that the belief in an unavoidable karma prompts one to despair because, despite one's noble determinations and aspirations, one knows of the canker within, and the likelihood of the necessity of many rounds of rebirth before one's inner baseness can be wholly purged away. At any rate, we do not have much confidence that the karma principle will explain the kind of pessimism we are trying to describe, which, we believe, is in fact a very simple, universal human response to human suffering, based partly on a type of primitive empathy, sympathy, putting myself in the other's place—of which all but the basest of mankind have some modicum, and against which, if they are human, all make some sort of protest.

Buddhism in time developed its own psychological system, a series of conceptions regarding the mind, the emotions, and the will, which we shall try briefly to make clear. The system differs from Hinduism most strikingly in the denial of a changeless inner principle. For the Buddhist it is the mental state itself, the confused perception, the painful emotion that has to be dealt with. There is no personal atman lurking behind and waiting to reveal itself when the mental states are removed; rather, the mental states are the reality, and there is, therefore, no persisting soul through all the changes. The Buddhists were of course accused by many of the Hindu philosophers of being grossly inconsistent since, in spite of all this, they believe that there is fulfillment or repayment in subsequent reincarnations for all that has been lived through in each phase of life. To the Hindu this proved that there is a changeless self. It proved no such thing to the Buddhist.

But when we speak of the flight of Buddhism across the great mountains to the north (before A.D. 1000), we find ourselves passing into the utterly different cultural context of Chinese civilization. Chinese Buddhism arose, not in a vacuum, but in competition with an ancient philosophical-poetic system known as Taoism, the philosophy of Lao-tzu (fifth century B.C.), defining a "way" or law or principle, or a norm or reality, which underlies all the passing events of this life. Moreover, a very concrete, practical system of laws of living had been defined by the sage Confucius, who had lived at the same time as Lao-tzu and was only a few years younger than Gautama, the Buddha. Chinese Buddhism takes on some of the trappings of Taoism. At the same time the intensely practical demands of the Chinese social order and family system led to a psychology which is really more ethics than it is psychology; a system of rules, obligations, practical principles which define the right, the natural, the inevitable outcome of this or that line of conduct. Chinese psychology has the intensely practical flavor of the Buddhist noble eightfold path, without much of the concern, as Dr. Francis Hsu says (page 145), with "first and last things" which characterizes the Indians.

Buddhism was carried everywhere north and east, and in time went on from China to Korea and Japan. Chinese characters, and therefore Chinese literature, and therefore in time Chinese religion, philosophy, and psychology all made their way into Japan, beginning in the sixth century of the Christian era. Buddhism took hold strongly in Japan, though emperors both in China and in Japan might regard it as dangerous because it led people at times to escape into religion: "They think too much about a life beyond death." Buddhism was a profound comfort and a practical guide not only to the lower strata of society but (as was also true of Christianity) to the middle and upper strata as well. Despite the continuation of the local and traditional Japanese religion of the Shinto shrines, Buddhism became a universal force. Its conception of practicalities and of right and wrong combined with its conception of a natural order and an ultimate escape from suffering. Buddhism, as a matter of fact, took various forms, some of which were peculiarly psychological, mystical, esoteric (pages 181 ff.). Zen Buddhism is one of these. As in the case of the archer drawing his arrow and letting fly at the target, there is a sort of ultimate unification of the mind with the object. This principle articulates closely with a whole series of Western conceptions of the process of per-

10

ceiving and the ways in which mind can empty itself and fill itself with some reality or event in the outer world.

These Asian systems, the systems of India, China, and Japan, are all both mystical and practical; they all begin with religion as directed toward immediate freedom from suffering and the ultimate salvation from sufferings otherwise built into the very nature of life. They all become psychological because in the course of defining religious and ethical practices they become more and more explicit as to what the mind is and how it works, how it can be freed both of delusion and of self-imposed misery.

*One*

# THE PSYCHOLOGY OF

# INDIA

## INTRODUCTION

. . . The religions of antiquity must always be approached in a loving spirit. . . . To watch in the *Sacred Books of the East* the dawn of the religious consciousness of man, must always remain one of the most inspiring and hallowing sights in the whole history of the world . . . What we want here, as everywhere else, is the truth, and the whole truth; and if the whole truth must be told, it is that, however radiant the dawn of religious thought, it is not without its dark clouds, its chilling colds, its noxious vapors. Whoever does not know these, or would hide them from his own sight and from the sight of others, does not know and can never understand the human heart in its first religious aspirations; and not knowing its toil and travail, can never know the intensity of its triumphs and its joys.*

Some preliterate men passed into the literate phase, and developed and systematized the art of writing, as did the Egyptians. They learned to keep records, to construct repositories of their own doings and thoughts. By 1500 B.C., the contents of their wastebaskets made up mountains of evidence for the modern Egyptologists to decipher and interpret. Yet they developed little or no psychology. Indeed, psychology is not the universal invention of all literate peoples. The first great giant step beyond preliterate psychology is the psychology of India.

How did India come to produce a psychology? It would be an interesting question to ask why some peoples and cultural epochs

* *The Sacred Books of the East,* edited by F. Max Müller (Oxford: Clarendon Press, 1879), I, xi.

15

have developed a psychology and some have not. It seems to us that there are two great steps that have to be taken before a people can develop a psychology. First, they have to have a rich, productive system, a surplus and accumulation of material goods, some leisure in which to carry out a sustained artistic or speculative effort, and a class of persons with the necessary freedom, independence, and detachment to develop an interest in speculative, far-reaching questions. This is to say that there first has to be a civilization which can allow itself a philosophy. Secondly, within the broad, far-flung empire of speculative thought, there has to be a specific concern with human nature—its essence and its doings.

At least, historically, this is the way it appears to us. There have been only a few independently arising civilizations. Most of them, in fact, have sprung up in tropical and subtropical areas and in great river valleys, where, as Turner has put it, there was an opportunity to "keep the jungle down." There was a chance for development of agricultural and industrial skills, commerce and credit, travel, and, at times, invasion, all adding to the variety of ideas, interacting with one another and forming a "culture base" on which a pyramid of new ideas could be erected. Of the half dozen or so great civilizations which arose under these conditions, two— India and China—developed philosophies from which psychologies could be derived. One more—that of the Greeks—came into being in the Mediterranean area. This likewise arose from a rich interaction of several subtropical civilizations (compare page 18). Civilization has always depended much on borrowing, on commerce and conquest, and on the rich fertilization of many ideas by ideas of many origins. We cannot very greatly advance knowledge of the origins of psychology by speculations of this sort, but at least we can note that it was under conditions of tropical wealth, leisure, and the hybridization of ideas from many sources that the first great Indian philosophies and, from them, the first great Indian psychologies arose.

It may be useful, first of all, to attempt a broad sweep, a geographical and cultural vista of the world in which the psychologies of India took shape. The Indian subcontinent south of the Himalayas is a practically unbroken plain with its high plateau of the

Deccan, its gentle rivers and their many tributaries, its openness to the starry skies for ten months of each year, the period between the great monsoons. No sharp subdivision—either of space or of time—is relevant to the slow and ceaseless rhythms of tropical living. The universe is in the palm of one's hand, and eternity is there in a respiratory cycle. Contrast the life of India with that of Greece, the tumultuous Greek mountains chopping up the land as the islands of the Aegean likewise represent little choppy portions of life with discontinuity from one to another, and with the sweep of maritime conquest alternating with invasion and subdivision of the soil into sharply accented portions of space which—with the staccato rhythm of life—means also subdivisions in time. In this respect, Greece is like all of the West, for there is no "great plain" in all of western Europe. For a parallel to India one would have to go to the other great Asian land mass north of the Himalayas, that which includes China and large parts of Siberia. China has always been, in ancient times as well as today, a geographical and a socio-cultural unity which, despite modulations of dialect and culture, has had a sense of inner continuity and has been regarded by the rest of the world as a relatively homogeneous portion of the globe. Just as India and China have contrasting political systems, so throughout all of history each has had a relative internal unity and they have confronted each other across the mountains as very different civilizations. As Francis Hsu has brilliantly shown (pages 174 ff.), there are profound religious and socio-cultural differences between India and China, so that the practical, matter-of-fact, socially regulated norms of Chinese life stand in stark opposition to the mystical, supernatural life patterns of the Indian subcontinent.

We do not know whether India presented a culturally homogeneous face to the world before the Aryan-speaking people came in from Persia in the second millennium before Christ. We do know, however, from the diggings in western India and what is now Pakistan, that a high civilization existed with a well-structured government, magnificent handicraft, and evidently a high state of specialized production and exchange involving commerce with Mesopotamia and lands farther to the west. We do know that mysticism and melancholy, rather than the high practicality of the

Olympian Greeks and the Chinese emperors, were conspicuous in the pre-Aryan religion (see pages 5 ff.). There may be a parallel with one of the thoughts offered by Gilbert Murray to the effect that the magnificent "swashbuckling" Olympian-worshiping Greeks ran full tilt into a thoughtful, troubled, melancholy, and mystical Aegean people and that the interaction of the two produced Greek civilization as we know it.

Nietzsche, in *The Birth of Tragedy*, had contrasted the "Apollonian" and "Dionysian" components in Greek life. Here we are, of course, allowing ourselves high and wide speculation which may need, over the years, to be refined. But we believe that a very gross and large parallel of a confident military people encountering a somber and melancholy native population may be drawn between Greek history and Indian history.

In any event, the high civilization which produced the tremendous philosophy and psychology of India took shape, not among the simple people who sang the Vedic hymns, but among their descendants, who were already mixing both biologically and culturally with the dark Dravidian peoples whom they slowly subjugated and brought within their own orbit of thought.

The religion of the Rig-Veda is, for the most part, a nature worship in which various forces of sun, sky, fire, water, and the sacrificial beverage have a personal or "animistic" character. The gods are persons, but they are also the forces of nature. There is nothing to suggest a mystical loss of individuality in the worshiper or in the deity. Something in the interaction with the tropic land or the primordial dark mysticism of the native peoples, or both, washed away the high confidence and primitive self-assertion of the warrior hosts. The priests began to standardize and ritualize the sacrifices and ceremonies while the warriors took over the speculative philosophizing functions. This took, of course, some hundreds of years. It was going on apace during the first millennium before Christ. Attention turned inward. The real problems were the problems of self-awareness, and ultimately of a kind of inner existence which went beyond the self and betokened a relation to the steady eternal reality of cosmic spirit, with which the individual spirit ultimately becomes united.

It is easy for us to become confused about these matters of spirituality and practicality, especially in India. Here is a vast population struggling to keep itself alive, developing thousands of specialized skills which are expressed in the caste organization; coping daily with the host of problems associated with poverty, insufficiency of soil and water, with the inevitable reliance on superstition and magic as devices to make tomorrow better than today; using prayer and superstitious devices; forced by abject poverty to a toughness and hardness toward those of other castes. The caste is, in some ways, like a family. Those within one's caste are one's blood brothers, those on whom one leans for favors and advancement. The upper castes have relied for millennia on the firmness of the caste barriers to give them wealth, power, and prestige, and it is, of course, among the elite Brahman caste that the great tradition is maintained most vigorously, as it is they to whom one turns as the priestly caste and for the maintenance of the right relations with the invisible world. Everywhere in India the Brahman is known as the vehicle of the great Hindu tradition. It is naturally they, the Brahmans, to whom intellectual privileges come; they are those who read and write; they administer both the civil and the religious authority. Indeed, the British, upon their arrival in southern India in the eighteenth century, assigned responsible middle-level positions in the empire largely to the Brahmans.

When, therefore, one speaks of the religious philosophy and psychology of India, one means in the first instance that of the Brahmans, bearing in mind, however, that as an elite the Brahmans always have something to offer to those less privileged. In northern India, where the Sanskrit language of the Aryan conquerors is the root of the vernacular tongues spoken by some three hundred million people, there is, of course, an intimate relation between the religious system of today and the great Sanskrit system in which the religious philosophy took shape two thousand years ago. Even in southern India, however, where the vernacular tongues are not derived from Sanskrit, the religious system and the caste system made their way just as they did in the north. The people of the non-Sanskrit language groups of southern India, numbering a hundred million or more, are just as good Hindus as those of the Sanskrit-derived linguistic groups of the north.

The main texts we shall use to illustrate the psychology of India are (1) the Rig-Veda, (2) the Upanishads, (3) the Bhagavad-Gita, (4) the Yoga sutras of Patanjali, (5) the Sermons of Buddha. This does not do justice to the whole of Indian psychology, but we thought it better to give considerable material on a few main sources than a large number of snippets covering dozens of psychological systems.

From our own experience in working with the ancient Indian sources, and from the experience of many other people, we are convinced that an overall introduction to Indian life, thought, and religion, written by an expert, must precede the specific extracts from the ancient sources. Such an introduction to Indian religion, philosophy, and psychology is available in the systematic work of C. E. Eliot (a scholar who served a period as His Majesty's Ambassador at Tokyo), in a volume, *Hinduism and Buddhism,* which, over the years, has given us a sense of seeing through Western eyes and, at the same time, seeing in terms of universal human issues. After giving Eliot's perspective, we shall present an extract from Heinrich Zimmer's *Philosophies of India,* as edited by Joseph Campbell. Some readers may prefer to leap forward immediately to the selections from the Rig-Veda, the Upanishads, the Bhagavad-Gita, the Yoga sutras of Pantanjali, and the Sermons of Buddha. For most readers, though, we suggest straightaway reading, for which the scholarly perspective of C. E. Eliot may give the first preparation for a more intimate understanding of the extracts from the classics.

## HINDUISM AND BUDDHISM*

### Rebirth and the Nature of the Soul

The most characteristic doctrine of Indian religion—rarely absent in India and imported by Buddhism into all the countries which it in-

* From C. E. Eliot, *Hinduism and Buddhism: An Historical Sketch,* 3 vols. (London: Longmans, Green, 1921, 1925), I, l–lxxvii.

fluenced—is that called metempsychosis, the transmigration of the soul or reincarnation. The last of these terms best expresses Indian, especially Buddhist, ideas, but still the usual Sanskrit equivalent, *samsara*, means migration. The body breaks up at death, but something passes on and migrates to another equally transitory tenement. Neither Brahmans nor Buddhists seem to contemplate the possibility that the human soul may be a temporary manifestation of the Eternal Spirit which comes to an end at death—a leaf on a tree or a momentary ripple on the water. It is always regarded as passing through many births, a wave traversing the ocean.

Hindu speculation has never passed through the materialistic phase, and the doctrine that the soul is annihilated at death is extremely rare in India. Even rarer perhaps is the doctrine that it usually enters on a permanent existence, happy or otherwise. The idea underlying the transmigration theory is that every state which we call existence must come to an end. If the soul can be isolated from all the accidents and accessories attaching to it, then there may be a state of permanence and peace but not a state comparable with human existence, however enlarged and glorified. But why does not this conviction of impermanence lead to the simpler conclusion that the end of physical life is the end of all life? Because the Hindus have an equally strong conviction of continuity: everything passes away and changes, but it is not true to say of anything that it arises from nothing or passes into nothing. If human organisms (or any other organisms) are mere machines, if there is nothing more to be said about a corpse than about a smashed watch, then (the Hindu thinks) the universe is not continuous. Its continuity means for him that there is something which eternally manifests itself in perishable forms but does not perish with them any more than water when a pitcher is broken or fire that passes from the wood it has consumed to fresh fuel.

These metaphors suggest that the doctrine of transmigration or reincarnation does not promise what we call personal immortality. I confess that I cannot understand how there can be personality in the ordinary human sense without a body. When we think of a friend, we think of a body and a character, thoughts and feelings, all of them connected with that body and many of them conditioned by it. But the immortal soul is commonly esteemed to be something equally present in a newborn babe, a youth, and an old man. If so, it cannot be a personality in the ordinary sense, for no one could recognize the spirit of a departed friend if it is something which was present in him the day he was born and different from all the characteristics which he acquired during life. The belief that we shall recognize our friends in another world assumes that these characteristics are immortal, but it is hard to understand how they can be so, especially as it is also assumed that there is nothing immortal in a dog, which possesses affection and intelligence, but that there is something immortal in a newborn infant which cannot be said to possess either.

In one way metempsychosis raises insuperable difficulties to the survival of personality, for if you become someone else, especially an animal, you are no longer yourself according to any ordinary use of language. But one of the principal forms taken by the doctrine in India makes a modified survival intelligible. For it is held that a newborn child brings with it as a result of actions done in previous lives certain predispositions and these after being developed and modified in the course of that child's life are transmitted to its next existence.

As to the method of transmission there are various theories, for in India the belief in reincarnation is not so much a dogma as an instinct innate in all and only occasionally justified by philosophers, not because it was disputed but because they felt bound to show that their own systems were compatible with it. One explanation is that given by the Vedanta philosophy, according to which the soul is accompanied in its migrations by the *sukshmasarira*, or subtle body, a counterpart of the mortal body but transparent and invisible, though material. The truth of this theory, as of all theories respecting ghosts and spirits, seems to me a matter for experimental verification, but the Vedanta recognizes that in our experience a personal individual existence is always connected with a physical substratum.

The Buddhist theory of rebirth is somewhat different, for Buddhism even in its later divagations rarely ceased to profess belief in Gautama's doctrine that there is no such thing as a soul—by which is meant no such thing as a permanent unchanging self, or atman. Buddhists are concerned to show that transmigration is not inconsistent with this denial of the atman. The ordinary and indeed inevitable translation of this word by "soul" leads to misunderstanding, for we naturally interpret it as meaning that there is nothing which survives the death of the body and a fortiori nothing to transmigrate. But in reality the denial of the atman applies to the living rather than to the dead. It means that in a living man there is no permanent, unchangeable entity, but only a series of mental states, and since human beings, although they have no atman, certainly exist in this present life, the absence of the atman is not in itself an obstacle to belief in a similar life after death or before birth. Infancy, youth, age, and the state immediately after death may form a series of which the last two are as intimately connected as any other two. The Buddhist teaching is that when men die in whom the desire for another life exists—as it exists in all except saints—then desire, which is really the creator of the world, fashions another being, conditioned by the character and merits of the being which has just come to an end. Life is like fire: its very nature is to burn its fuel. When one body dies, it is as if one piece of fuel were burned: the vital process passes on and recommences in another, and, so long as there is desire of life, the provision of fuel fails not. Buddhist doctors have busied themselves with the question whether two successive lives are the same man or different men, and have illustrated the relationship by various analogies of things which seem to be the same and yet not the same, such as a child and an adult, milk and curds, or fire which spreads from a lamp and burns down a village, but, like the Brahmans, they do not discuss why the hypothesis of transmigration is necessary. They had the same feeling for the continuity of nature, and more than others they insisted on the principle that everything has a cause. They held that the sexual act creates the conditions in which a new life appears but is not an adequate cause for the new life itself. And unless we accept a materialist explanation of human nature, this argument is sound: unless we admit that mind is merely a function of matter, the birth of a mind is not explicable as a mere process of cell development: something pre-existent must act upon the cells.

Europeans, in discussing such questions as the nature of the soul and immortality, are prone to concentrate their attention on death and neglect the phenomena of birth, which surely are equally important. For if a soul survives the death of this complex of cells which is called the body, its origin and development must, according to all analogy, be different from those of the perishable body. Orthodox theology deals with the problem by saying that God creates a new soul every time a child is born, but free discussion usually ignores it and, taking an adult as he is, asks what are the chances that any part of him survives death. Yet the questions what is destroyed at death and how and why, are closely connected with the questions what comes into existence at birth and how and why. This second series of questions is hard enough, but it has this advantage over the first that whereas death abruptly closes the road and we cannot follow the soul one inch on its journey beyond, the portals of birth are a less absolute frontier. We know that every child has passed through stages in which it could hardly be called a child. The earliest phase consists of two cells, which unite and then proceed to subdivide and grow. The mystery of the process by which they assume a human form is not explained by scientific or theological phrases. The complete individual is assuredly not contained in the first germ. The microscope cannot find it there and to say that it is there potentially merely means that we know the germ will develop in a certain way. To say that a force is manifesting itself in the germ and assuming the shape which it chooses to take or must take is also merely a phrase and metaphor, but it seems to me to fit the facts.

The doctrines of pre-existence and transmigration (but not, I think, of karma, which is purely Indian) are common among savages in Africa and America; nor is their wide distribution strange. Savages commonly think that the soul wanders during sleep and that a dead man's soul goes somewhere: what more natural than to suppose that the soul or a new-born infant comes from somewhere? But among civilized peoples such ideas are in most cases due to Indian influence. In India they seem indigenous to the soil and not imported by the Aryan invaders, for they are not clearly enunciated in the Rig-Veda, nor formulated before the time of the Upanishads. They were introduced by Buddhism to the Far East and their presence in Manichaeism, Neoplatonism, Sufism, and ultimately in the Jewish Kabbala seems a rivulet from the same source. Recent research discredits the theory that metempsychosis was an important feature in the earlier religion of Egypt or among the Druids. But it played a prominent part in the philosophy of Pythagoras and in the Orphic mysteries, which had some connection with Thrace and possibly also with Crete. A few great European intellects—notably Plato and Virgil—have given it undying expression, but Europeans as a whole have rejected it with that curiously crude contempt which they have shown until recently for Oriental art and literature.

Considering how fixed is the belief in immortality among Europeans, or at least the desire for it, the rarity of a belief in pre-existence or transmigration is remarkable. But most people's expectation of a future life is based on craving rather than on reasoned anticipation. I cannot myself understand how anything that comes into being can be immortal. Such immortality is unsupported by a single analogy nor can any instance be quoted of a thing which is known to have had an origin and yet is even apparently indestructible. And is it possible to suppose that the universe is capable of indefinite increase by the continual addition of new and eternal souls? But these difficulties do not exist for theories which regard the soul as something existing before as well as after the body, truly immortal *a parte ante* as well as *a parte post* and manifesting itself in temporary homes of human or lower shape. Such theories become very various and fall into many obscurities when they try to define the nature of the soul and its relation to the body, but they avoid what seems to me the contradiction of the created but immortal soul.

The doctrine of metempsychosis is also interesting as affecting the relations of men and animals. The popular European conception of "the beasts which perish" weakens the arguments for human immortality. For if the mind of a dog or chimpanzee contains no element which is immortal, the part of the human mind on which the claim to immortality can be based must be parlously small, since *ex hypothesi* sensation, volition, desire, and the simpler forms of intelligence are not immortal.

But in India, where men have more charity and more philosophy, this distinction is not drawn. The animating principle of men, animals, and plants is regarded as one or at least similar, and even matter which we consider inanimate, such as water, is often considered to possess a soul. But though there is ample warrant in both Brahmanic and Buddhist literature for the idea that the soul may sink from a human to an animal form or vice versa rise, and, though one sometimes meets this belief in modern life, yet it is not the most prominent aspect of metempsychosis in India and the beautiful precept of *ahimsa*, or not injuring living things, is not, as Europeans imagine, founded on the fear of eating one's grand-parents, but rather on the humane and enlightened feeling that all life is one and that men who devour beasts are not much above the level of the beasts who devour one another. The feeling has grown stronger with time. In the Vedas animal sacrifices are prescribed and they are even now used in the worship of some deities. In the Epics the eating of meat is mentioned. But the doctrine that it is wrong to take animal life was definitely adopted by Buddhism and gained strength with its diffusion.

One obvious objection to all theories of rebirth is that we do not remember our previous existences and that, if they are connected by no thread of memory, they are for all practical purposes the existences of different people. But this want of memory affects not only past existences but the early phases of this existence. Does any one deny his existence as an infant or embryo because he cannot remember it? And if a wrong could be done to an infant the effects of which would not be felt for twenty years, could it be said to be no concern of the infant because the person who will suffer in twenty years time will have no recollection that he was that infant? And common opinion in Eastern Asia, not without occasional confirmation from Europe, denies the proposition that we cannot remember our former lives and asserts that those who take any pains to sharpen their spiritual faculties can remember them. The evidence for such recollection seems to me better than the evidence for most spiritualistic phenomena.

Another objection comes from the facts of heredity. On the whole we resemble our parents and ancestors in mind as well as in body. A child often seems to be an obvious product of its parents and not a being come from outside and from another life. This objection of course applies equally to the creation theory. If the soul is created by an act of God, there seems to be no reason why it should be like the parents, or, if he causes it to be like them, he is made responsible for sending children into the world with vicious natures. On the other hand, if parents literally make a child, mind as well as body, there seems to be no reason why children should ever be unlike their parents, or brothers and sisters un-like one another, as they undoubtedly sometimes are. An Indian would say that a soul seeking rebirth carries with it certain potentialities of good and evil and can obtain embodiment only in a family offering the necessary conditions. Hence to some extent it is natural that the child should be like its parents. But the soul seeking rebirth is not completely fixed in form and stiff: it is hampered and limited by the results of its previous life, but in many respects it may be flexible and free, ready to vary in response to its new environment.

But there is a psychological and temperamental objection to the doctrine of rebirth, which goes to the root of the matter. Love of life and the desire to find a field of activity are so strong in most Europeans that it might be supposed that a theory offering an endless vista of new activities and new chances would be acceptable. But as a rule Europeans who discuss the question say that they do not relish this prospect. They may be willing to struggle until death, but they wish for repose—conscious repose of course—afterward. The idea that one just dead has not entered into his rest, but is beginning another life with similar struggles and fleeting successes, similar sorrows and disappointments, is not satisfying and is almost shocking. We do not like it, and not to like any particular view about the destinies of the soul is generally, but most illogically, considered a reason for rejecting it.

It must not, however, be supposed that Hindus like the prospect of transmigration. On the contrary, from the time of the Upanishads and the Buddha to the present day, their religious ideal corresponding to salvation is emancipation and deliverance, deliverance from rebirth and from the bondage of desire which brings about rebirth. Now all Indian theories as to the nature of transmigration are in some way connected with the idea of karma, that is the power of deeds done in past existences to condition or even to create future existences. Every deed done, whether good or bad, affects the character of the doer for a long while, so that, to use a metaphor, the soul awaiting rebirth has a special shape, which is of its own making, and it can find re-embodiment only in a form into which that shape can squeeze.

These views of rebirth and karma have a moral value, for they teach that what a man gets depends on what he is or makes himself to be, and they avoid the difficulty of supposing that a benevolent creator can have given his creatures only one life with such strange and unmerited disproportion in their lots. Ordinary folk in the East hope that a life of virtue will secure them another life as happy beings on earth or perhaps in some heaven which, though not eternal, will still be long. But for many the higher ideal is renunciation of the world and a life of contemplative asceticism which will accumulate no karma so that after death the soul will pass not to another birth but to some higher and more mysterious state which is beyond birth and death. It is the prevalence of views like this which has given both Hinduism and Buddhism the reputation of being pessimistic and unpractical.

It is generally assumed that these are bad epithets, but are they not applicable to Christian teaching? Modern and medieval Christianity—as witness many popular hymns—regards this world as vain and transitory, a vale of tears and tribulation, a troubled sea through whose waves we must pass before we reach our rest. And choirs sing, though without much conviction, that it is weary waiting here. This language seems justified by the Gospels and Epistles. It is true that some utterances of Christ suggest that happiness is to be found in a simple and natural life of friendliness and love, but on the whole both he and St. Paul teach that the world is evil or at least spoiled and distorted: to become a happy world it must be somehow remade and transfigured by the second coming of Christ. The desires and ambitions which are the motive power of modern Europe are, if not wrong, at least vain and do not even seek for true peace and happiness. Like Indian teachers, the early Christians tried to create a right temper rather than to change social institutions. They bade masters and slaves treat one another with kindness and respect, but they did not attempt to abolish slavery.

Indian thought does not really go much further in pessimism than Christianity, but its pessimism is intellectual rather than emotional. He who understands the nature of the soul and its successive lives cannot regard any single life as of great importance in itself, though its consequences for the future may be momentous, and though he will not say that life is not worth living. Reiterated declarations that all existence is suffering do, it is true, seem to destroy all prospect of happiness and all motive for effort, but the more accurate statement is, in the words of the Buddha himself, that all clinging to physical existence involves suffering. The earliest Buddhist texts teach that when this clinging and craving cease, a feeling of freedom and happiness takes their place and later Buddhism treated itself to visions of paradise as freely as Christianity. Many forms of Hinduism teach that the soul released from the body can enjoy eternal bliss in the presence of God and even those severer philosophers who do not admit that the released soul is a personality in any human sense have no doubt of its happiness.

The opposition is not so much between Indian thought and the New Testament, for both of them teach that bliss is attainable but not by satisfying desire. The fundamental contrast is rather between both India and the New Testament on the one hand and on the other the rooted conviction of European races, however much Christian orthodoxy may disguise their expression of it, that this world is all-important. This conviction finds expression not only in the avowed pursuit of pleasure and ambition but in such sayings as that the best religion is the one which does most good and such ideals as self-realization or the full development of one's nature and powers. Europeans as a rule have an innate dislike and mistrust of the doctrine that the world is vain or unreal. They can accord some sympathy to a dying man who sees in due perspective the unimportance of his past life or to a poet who under the starry heavens can make felt the smallness of man and his earth. But such thoughts are considered permissible only as retrospects, not as principles of life: you may say that your labor has amounted to nothing, but not that labor is vain. Though monasteries and monks still exist, the great majority of Europeans instinctively disbelieve in asceticism, the contemplative life, and contempt of the world: they have no love for a philosopher who rejects the idea of progress and is not satisfied with an ideal consisting in movement toward an unknown goal. They demand a religion which theoretically justifies the strenuous life. All this is a matter of temperament and the temperament is so common that it needs no explanation. What needs explanation is rather the other temperament which rejects this world as unsatisfactory and sets up another ideal, another sphere, another standard of values. This ideal and standard are not entirely peculiar to India but certainly they are understood and honored there more than elsewhere. They are professed, as I have already observed, by Christianity, but even the New Testament is not free from the idea that saints are having a bad time now but will hereafter enjoy a triumph, parlously like the exuberance of the wicked in this world. The Far East too has its unworldly side which, though harmonizing with Buddhism, is native. In many ways the Chinese are as materialistic as Europeans, but throughout the long history of their art and literature there has always been a school, clear-voiced if small, which has sung and pursued the joys of the hermit, the dweller among trees and mountains who finds nature and his own thoughts an all-sufficient source of continual happiness. But the Indian ideal, though it often includes the pleasures of communion with nature, differs from most forms of the Chinese and Christian ideal inasmuch as it assumes the reality of certain religious experiences and treats them as the substance and occupation of the highest life. We are disposed to describe these experiences as trances or visions, names which generally mean something morbid or hypnotic. But in India their validity is unquestioned and they are not considered morbid. The sensual scheming life of the world is sick and ailing; the rapture of contemplation is the true and healthy life of the soul. More than that, it is the type and foretaste of a higher existence compared with which this world is worthless or rather nothing at all. This view has been held in India for nearly three thousand years: it has been confirmed by the experience of men whose writings testify to their intellectual power and

has commanded the respect of the masses. It must ⌐
too, even if it is contrary to our temperament, for it
of a great nation and cannot be explained awa⅄
charlatanism. It is allied to the experiences of
whom St. Teresa is a striking example, though le⅄
as Walt Whitman and J. A. Symonds, might ⅌
mysticism William James said, "The existence ⌐
lutely overthrows the pretension of non-mystical states ⅃⌐ ⌐
ultimate dictators of what we may believe."

These mystical states are commonly described as meditation, but they include not merely peaceful contemplation but ecstatic rapture. They are sometimes explained as union with Brahman, the absorption of the soul in God, or its feeling that it is one with him. But this is certainly not the only explanation of ecstasy given in India, for it is recognized as real and beneficent by Buddhists and Jains. The same rapture, the same sense of omniscience and of ability to comprehend the scheme of things, the same peace and freedom, are experienced by both theistic and non-theistic sects, just as they have also been experienced by Christian mystics. The experiences are real, but they do not depend on the presence of any special deity, though they may be colored by the theological views of individual thinkers. The earliest Buddhist texts make right rapture (*samma samadhi*) the end and crown of the eightfold path but offer no explanation of it. They suggest that it is something wrought by the mind for itself and without the cooperation or infusion of any external influence.

Indian ideas about the destiny of the soul are connected with equally important views about its nature. I will not presume to say what is the definition of the soul in European philosophy, but in the language of popular religion it undoubtedly means that which remains when a body is arbitrarily abstracted from a human personality, without inquiring how much of that personality is thinkable without a material substratum. This popular soul includes mind, perception, and desire and often no attempt is made to distinguish it from them. But in India it is so distinguished. The soul (atman, or *purusha*) *uses* the mind and senses: they are its instruments rather than parts of it. Sight, for instance, serves as the spectacles of the soul, and the other senses and even the mind (*manas*), which is an intellectual *organ*, are also instruments. If we talk of a soul passing from death to another birth, this according to most Hindus is a soul accompanied by its baggage of mind and senses, a subtle body indeed, but still gaseous not spiritual. But what is the soul by itself? When an English poet sings of death that it is "Only the sleep eternal in an eternal night" or a Greek poet calls it *atermona něgreton hypnon*, we

at they are denying immortality. But Indian divines maintain that
p sleep is one of the states in which the soul approaches nearest to
God: that it is a state of bliss and is unconscious not because consciousness is suspended but because no objects are presented to it. Even higher
than dreamless sleep is another condition known simply as the fourth
state (*turiya* or *caturtha*), the others being waking, dream-sleep, and
dreamless sleep. In this fourth state thought is one with the object of
thought and, knowledge being perfect, there exists no contrast between
knowledge and ignorance. All this sounds strange to modern Europe.
We are apt to say that dreamless sleep is simply unconsciousness and
that the so-called fourth state is imaginary or unmeaning. But to follow
even popular speculation in India it is necessary to grasp this truth, or
assumption, that when discursive thought ceases, when the mind and the
senses are no longer active, the result is not unconsciousness equivalent
to nonexistence but the highest and purest state of the soul, in which,
rising above thought and feeling, it enjoys the untrammeled bliss of its
own nature.

If these views sound mysterious and fanciful, I would ask those
Europeans who believe in the immortality of the soul what, in their
opinion, survives death. The brain, the nerves, and the sense organs
obviously decay: the soul, you may say, is not a product of them, but
when they are destroyed or even injured, perceptive and intellectual
processes are inhibited and apparently rendered impossible. Must not that
which lives forever be, as the Hindus think, independent of thought and
of sense impressions?

I have observed in my reading that European philosophers are more
ready to talk about soul and spirit than to define them and the same is
true of Indian philosophers. The word most commonly rendered by soul
is "atman," but no one definition can be given for it, for some hold that
the soul is identical with the Universal Spirit, others that it is merely
of the same nature, still others that there are innumerable souls uncreate
and eternal, while the Buddhists deny the existence of a soul in toto. But
most Hindus who believe in the existence of an atman or soul agree in
thinking that it is the real self and essence of all human beings (or for
that matter of other beings): that it is eternal *a parte ante* and *a parte
post:* that it is not subject to variation but passes unchanged from one
birth to another: that youth and age, joy and sorrow, and all the accidents of human life are affections, not so much of the soul as of the
envelopes and limitations which surround it during its pilgrimage: that
the soul, if it can be released and disengaged from these envelopes, is in
itself knowledge and bliss, knowledge meaning the immediate and intuitive knowledge of God. A proper comprehension of this point of view
will make us chary of labeling Indian thought as pessimistic on the

ground that it promises the soul something which we are inclined to call unconsciousness.

In studying Oriental religions, sympathy and a desire to agree if possible are the first requisites. For instance, he who says of a certain ideal, "This means annihilation and I do not like it," is on the wrong way. The right way is to ascertain what many of our most intelligent brothers mean by the cessation of mental activity and why it is for them an ideal.

## Eastern Pessimism and Renunciation

But the charge of pessimism against Eastern religions is so important that we must consider other aspects of it, for though the charge is wrong, it is wrong only because those who bring it do not use quite the right word. And indeed it would be hard to find the right word in a European language. The temperament and theory described as pessimism are European. They imply an attitude of revolt, a right to judge and grumble. Why did the Deity make something out of nothing? What was his object? But this is not the attitude of Eastern thought: it generally holds that we cannot imagine nothing: that the world process is without beginning or end and that man must learn how to make the best of it.

The Far East purged Buddhism of much of its pessimism. There we see that the First Truth about suffering is little more than an admission of the existence of evil, which all religions and common sense admit. Evil ceases in the saint: nirvana in this life is perfect happiness. And though striving for the material improvement of the world is not held up conspicuously as an ideal in the Buddhist scriptures (or for that matter in the New Testament), yet it is never hinted that good effort is vain. A king should be a good king.

Renunciation is a great word in the religions of both Europe and Asia, but in Europe it is almost active. Except to advanced mystics, it means abandoning a natural attitude and deliberately assuming another which it is difficult to maintain. Something similar is found in India in the legends of those ascetics who triumphed over the flesh until they become very gods in power. But it is also a common view in the East that he who renounces ambition and passion is not struggling against the world and the devil but simply leading a natural life. His passions indeed obey his will and do not wander here and there according to their fancy, but his temperament is one of acquiescence not resistance. He takes his place among the men, beasts, and plants around him and, ceasing to struggle, finds that his own soul contains happiness in itself.

Most Europeans consider man the center and lord of the world or, if they are very religious, its viceregent under God. He may kill or otherwise maltreat animals for his pleasure or convenience: his task is to subdue the forces of nature: nature is subservient to him and to his destinies: without man nature is meaningless. Much the same view was held by the ancient Greeks and in a less acute form by the Jews and Romans. Swinburne's line

Glory to man in the highest, for man is the master of things

is overbold for professing Christians but it expresses both the modern scientific sentiment and the ancient Hellenic sentiment.

But such a line of poetry would, I think, be impossible in India or in any country to the east of it. There man is thought of as a part of nature, not its center or master. Above him are formidable hosts of deities and spirits, and even European engineers cannot subdue the genii of the flood and typhoon: below but still not separated from him are the various tribes of birds and beasts. A good man does not kill them for pleasure nor eat flesh, and even those whose aspirations to virtue are modest treat animals as humble brethren rather than as lower creatures over whom they have dominion by divine command.

This attitude is illustrated by Chinese and Japanese art. In architecture, this art makes it a principle that palaces and temples should not dominate a landscape but fit into it and adapt their lines to its features. For the painter, flowers and animals form a sufficient picture by themselves and are not felt to be inadequate because man is absent. Portraits are frequent but a common form of European composition, namely a group of figures subordinated to a principal one, though not unknown, is comparatively rare.

How scanty are the records of great men in India! Great buildings attract attention, but who knows the names of the architects who planned them or the kings who paid for them? We are not quite sure of the date of Kalidasa, the Indian Shakespeare, and though the doctrines of Sankara, Kabir, and Nanak still flourish, it is with difficulty that the antiquary collects from the meager legends clinging to their names a few facts for their biographies. And kings and emperors, a class who in Europe can count on being remembered if not esteemed after death, fare even worse. The laborious research of Europeans has shown that Asoka and Harsha were great monarchs. Their own countrymen merely say, "Once upon a time there was a king," and recount some trivial story.

In fact, Hindus have a very weak historical sense. In this they are not wholly wrong, for Europeans undoubtedly exaggerate the historical treatment of thought and art. In science, most students want to know what is certain in theory and useful in practice, not what were the discarded hypotheses and imperfect instruments of the past. In literature, when the actors and audience are really interested, the date of Shakespeare and even the authorship of the play cease to be important. In the same way Hindus want to know whether doctrines and speculations are true, whether a man can make use of them in his own religious experiences and aspirations. They care little for the date, authorship, unity, and textual accuracy of the Bhagavad-Gita. They simply ask, is it true, what can I get from it? The European critic, who expects nothing of the sort from the work, racks his brains to know who wrote it and when, who touched it up and why?

The Hindus are also indifferent to the past because they do not recognize that the history of the world, the whole cosmic process, has any meaning or value. In most departments of Indian thought, great or small, the conception of *telos*, or purpose, is absent, and if the European reader thinks this a grave lacuna, let him ask himself whether satisfied love has any *telos*. For Hindus the world is endless repetition, not a progress toward an end. Creation has rarely the sense which it bears for Europeans. An infinite number of times the universe has collapsed in flaming or watery ruin, eons of quiescence follow the collapse, and then the Deity (he has done it an infinite number of times) emits again from himself worlds and souls of the same old kind. But though all varieties of theological opinion may be found in India, he is usually represented as moved by some reproductive impulse rather than as executing a plan. Sankara says boldly that no motive can be attributed to God, because he, being perfect, can desire no addition to his perfection, so that his creative activity is mere exuberance, like the sport of young princes, who take exercise though they are not obliged to do so.

Such views are distasteful to Europeans. Our vanity impels us to invent explanations of the universe which make our own existence important and significant. Nor does European science altogether support the Indian doctrine of periodicity. It has theories as to the probable origin of the solar system and other similar systems, but it points to the conclusion that the universe as a whole is not appreciably affected by the growth or decay of its parts, whereas Indian imagination thinks of universal cataclysms and recurring periods of quiescence in which nothing whatever remains except the undifferentiated divine spirit.

Western ethics generally aim at teaching a man how to act: Eastern ethics at forming a character. A good character will no doubt act rightly when circumstances require action, but he need not seek occasions for action, he may even avoid them, and in India the passionless sage is still, in popular esteem, superior to warriors, statesmen, and scientists.

## Eastern Polytheism

Different as India and China are, they agree in this that in order not to misapprehend their religious condition we must make our minds familiar with a new set of relations. The relations of religion to philosophy, to ethics, and to the state, as well as the relations of different religions to one another, are not the same as in Europe. China and India are pagan, a word which I deprecate if it is understood to imply inferiority but which if used in a descriptive and respectful sense is very useful. Christianity and Islam are organized religions. They say (or rather their several sects say) that they each not only possess the truth but that all other creeds and rites are wrong. But paganism is not organized: it rarely presents anything like a church united under one head: still more rarely does it condemn or interfere with other religions unless attacked first. Buddhism stands between the two classes. Like Christianity and Islam, it professes to teach the only true law, but, unlike them, it is exceedingly tolerant and many Buddhists also worship Hindu or Chinese gods.

Popular religion in India and China is certainly polytheistic, yet if one uses this word in contrast to the monotheism of Islam and of Protestantism, the antithesis is unjust, for the polytheist does not believe in many creators and rulers of the world, in many Allahs or Jehovahs, but he considers that there are many spiritual beings, with different spheres and powers, to the most appropriate of whom he addresses his petitions. Polytheism and image worship lie under an unmerited stigma in Europe. We generally assume that to believe in one God is obviously better, intellectually and ethically, than to believe in many. Yet Trinitarian religions escape being polytheistic only by juggling with words, and if Hindus and Chinese are polytheists so are the Roman and Oriental churches, for there is no real distinction between praying to the Madonna, saints, and angels, and propitiating minor deities. William James has pointed out that polytheism is not theoretically absurd and is practically the religion of many Europeans. In some ways it is more intelligible and reasonable than monotheism. For if there is only one personal God, I do not understand how anything that can be called a person can be so expanded as to be capable of hearing and answering the prayers of the whole world. Anything susceptible of such extension must be more than a person. Is it not at least equally reasonable to assume that there are many spirits, or many shapes taken by the superpersonal world spirit, with which the soul can get into touch?

The worship of images cannot be recommended without qualification, for it seems to require artists capable of making a worthy representation of the divine. And it must be confessed that many figures in Indian temples, such as the statutes of Kali, seem repulsive or grotesque, though a Hindu might say that none of them are so strange in idea or so horrible in appearance as the crucifix. But the claim of the iconoclast from the times of the Old Testament onward that he worships a spirit whereas others worship wood and stone is true only of the lowest phases of religion, if even there. Hindu theologians distinguish different kinds of *avataras*, or ways in which God descends into the world: among them are incarnations like Krishna, the presence of God in the human heart and his presence in a symbol or image (*arca*). It may be difficult to decide how far the symbol and the spirit are kept separate either in the East or in Europe, but no one can attend a great car festival in southern India or the feast of Durga in Bengal without feeling and in some measure sharing the ecstasy and enthusiasm of the crowd. It is an enthusiasm such as may be evoked in critical times by a king or a flag, and as the flag may do duty for the king and all that he stands for, so may the image do duty for the deity.

## The Extravagance of Hinduism

What I have just said applies to India rather than to China and so do the observations which follow. India is the most religious country in the world. The percentage of people who literally make religion their chief business, who sacrifice to it money and life itself (for religious suicide is not extinct), is far greater than elsewhere. Russia probably comes next but the other nations fall behind by a long interval. Matter-of-fact, respectable people—Chinese as well as Europeans—call this attitude extravagance and it sometimes deserves the name, for since there is no one creed or criterion in India, all sorts of aboriginal or decadent superstitions command the respect due to the name of religion.

This extravagance is both intellectual and moral. No story is too extraordinary to be told of Hindu gods. They are the magicians of the universe who sport with the forces of nature as easily as a conjuror in a bazaar does tricks with a handful of balls. But though the average Hindu would be shocked to hear the Puranas described as idle tales, yet he does not make his creed depend on their accuracy, as many in Europe make Christianity depend on miracles. The value of truth in religion is rated higher in India than in Europe, but it is not historical truth. The Hindu approaches his sacred literature somewhat in the spirit in which we approach Milton and Dante. The beauty and value of such poems is clear. The question whether they are accurate reports of facts seems irrelevant. Hindus believe in progressive revelation. Many Tantras and Vishnuite works profess to be better suited to the present age than the Vedas, and innumerable treatises in the vernacular are commonly accepted as scripture.

Scriptures in India are thought of as words not writings. It is the sacred sound not a sacred book which is venerated. They are learned by oral transmission and it is rare to see a book used in religious services. Diagrams accompanied by letters and a few words are credited with magical powers, but still tantric spells are things to be recited rather than written. This view of scripture makes the hearer uncritical. The ordinary layman hears parts of a sacred book recited and probably admires what he understands, but he has no means of judging of a book as a whole, especially of its coherency and consistency.

The moral extravagance of Hinduism is more serious. It is kept in check by the general conviction that asceticism, or at least temperance, charity, and self-effacement are the indispensable outward signs of religion, but still among the great religions of the world there is none which countenances so many hysterical, immoral, and cruel rites. A literary example will illustrate the position. It is taken from the drama Madhava and Malati written about A.D. 730, but the incidents of the plot might happen in any native state today. In it Madhava, a young Brahman, surprises a priest of the goddess Chamunda who is about to immolate Malati. He kills the priest, and apparently the other characters consider his conduct natural and not sacrilegious. But it is not suggested that either the police or any ecclesiastical authority ought to prevent human sacrifices, and the reason why Madhava was able to save his beloved from death was that he had gone to the uncanny spot where such rites were performed to make an offering of human flesh to demons.

In Buddhism religion and the moral law are identified, but not in Hinduism. Brahmanic literature contains beautiful moral sayings, especially about unselfishness and self-restraint, but the greatest popular gods such as Vishnu and Siva are not identified with the moral law. They are super-moral and the God of philosophy, who *is* all things, is also above good and evil. The aim of the philosophic saint is not so much to choose the good and eschew evil as to draw nearer to God by rising above both.

Indian literature as a whole has a strong ethical and didactic flavor, yet the great philosophic and religious systems concern themselves little with ethics. They discuss the nature of the external world and other metaphysical questions which seem to us hardly religious: they clearly feel a peculiar interest in defining the relation of the soul to God, but they rarely ask why should I be good or what is the sanction of morality. They are concerned less with sin than with ignorance: virtue is indispensable, but without knowledge it is useless. . . .

## Morality and Will

It is dangerous to make sweeping statements about the huge mass of Indian literature, but I think that most Buddhist and Brahmanic systems assume that morality is merely a means of obtaining happiness and is not obedience to a categorical imperative or to the will of God. Morality is by inference raised to the status of a cosmic law, because evil deeds will infallibly bring evil consequences to the doer in this life or in another. But it is not commonly spoken of as such a law. The usual point of view is that man desires happiness and for this morality is a necessary though insufficient preparation. But there may be higher states which cannot be expressed in terms of happiness.

The will receives more attention in European philosophy than in Indian, whether Buddhist or Brahmanic, which both regard it not as a separate kind of activity but as a form of thought. As such it is not neglected in Buddhist psychology: will, desire, and struggle are recognized as good provided their object is good, a point overlooked by those who accuse Buddhism of preaching inaction.

Schopenhauer's doctrine that will is the essential fact in the universe and in life may appear to have analogies to Indian thought: it would be easy for instance to quote passages from the Pitakas showing that *tanha* (thirst, craving, or desire) is the force which makes and remakes the world. But such statements must be taken as generalizations respecting the wcrld as it is rather than as implying theories of its origin, for though *tanha* is a link in the chain of causation, it is not regarded as an ultimate principle more than any other link but is made to depend on feeling. The *maya* of the Vedanta is not so much the affirmation of the will to live as the illusion that we have a real existence apart from Brahman, and the same may be said of *ahamkara* in the Sankhya philosophy. It is the principle of egoism and individuality, but its essence is not so much self-assertion as the *mistaken* idea that this is *mine*, that *I* am happy or unhappy.

There is a question much debated in European philosophy but little argued in India, namely the freedom of the will. The active European, feeling the obligation and the difficulties of morality, is perplexed by the doubt whether he really has the power to act as he wishes. This problem has not much troubled the Hindus and rightly, as I think. For if the human will is not free, what does freedom mean? What example of freedom can be quoted with which to contrast the supposed non-freedom of the will? If in fact it is from the will that our notion of freedom is derived, is it not unreasonable to say that the will is not free? Absolute freedom in the sense of something regulated by no laws is unthinkable. When a thing is conditioned by external causes it is dependent. When it is conditioned by internal causes which are part of its own nature, it is free. No other freedom is known. An Indian would say that a man's nature is limited by karma. Some minds are incapable of the higher forms of virtue and wisdom, just as some bodies are incapable of athletic feats. But within the limits of his own nature a human being is free. Indian theology is not much hampered by the mad doctrine that God has predestined some souls to damnation, nor by the idea of fate, except insofar as karma is fate. It is fate in the sense that karma inherited from a previous birth is a store of rewards and punishments which must be enjoyed or endured, but it differs from fate because we are all the time making our own karma and determining the character of our next birth.

The older Upanishads hint at a doctrine analogous to that of Kant, namely that man is bound and conditioned insofar as he is a part of the world of phenomena but free insofar as the self within him is identical with the divine self which is the creator of all bonds and conditions. Thus the Kaushitaki Upanishad says, "He it is who causes the man whom he will lead upward from these worlds to do good works and He it is who causes the man whom he will lead downward to do evil works. He is the guardian of the world, He is the ruler of the world, He is the Lord of the world and He is myself." Here the last words destroy the apparent determinism of the first part of the sentence. And similarly the Chandogya Upanishad says, "They who depart hence without having known the Self and those true desires, for them there is no freedom in all worlds. But they who depart hence after knowing the Self and those true desires, for them there is freedom in all worlds."

Early Buddhist literature asserts uncompromisingly that every state of consciousness has a cause and in one of his earliest discourses the Buddha argues that the Skandhas, including mental states, cannot be the Self because we have not free will to make them exactly what we choose. But throughout his ethical teaching, it is, I think, assumed that, subject to the law of karma, conscious action is equivalent to spontaneous action. Good mental states can be made to grow and bad mental states to decrease until the stage is reached when the saint knows that he is free. It may perhaps be thought that the early Buddhists did not realize the consequences of applying their doctrine of causation to psychology and hence never faced the possibility of determinism. But determinism, fatalism, and the uselessness of effort formed part of the paradoxical teaching of Makkhali Gosala reported in the Pitakas and therefore well known. If neither the Jains nor the Buddhists allowed themselves to be embarrassed by such denials of free will, the inference is that in some matters at least the Hindus had strong common sense and declined to accept any view which takes away from man the responsibility and lordship of his own soul.

38

In cosmic terms, order, truth, and justice depend on right relationships of all the parts to one another and to the whole. Right
conduct is not only the chord by which the individual responds to
a cosmic symphony: it *creates* the symphony. Man and woman act
in accord with truth; and truth alters the world. Psychology is a
sort of rhythm of soul to world; and a rhythm which changes the
world. This is dramatically expressed in the story of the Ganges
flowing backward. This story, from the Milindapanha (A.D. 200),
is used in the following passage from Heinrich Zimmer's *Philosophies of India* to illuminate the Satya philosophy so characteristic
of Indian thought:

### SATYA*

The story is told, for example, of a time when the righteous King
Asoka, greatest of the great northern Indian dynasty of the Mauryas,
"stood in the city of Pataliputra, surrounded by the city folk and country folk, by his ministers and his army and his councilors, with the
Ganges flowing by, filled up by freshets, level with the banks, full to the
brim, five hundred leagues in length, a league in breadth. Beholding the
river, he said to his ministers, 'Is there anyone who can make this mighty
Ganges flow back upstream?' To which the ministers replied, 'That is a
hard matter, Your Majesty.'

"Now there stood on that very riverbank an old courtesan named
Bindumati, and when she heard the king's question she said, 'As for
me, I am a courtesan in the city of Pataliputra. I live by my beauty; my
means of subsistence is the lowest. Let the king but behold my Act of
Truth.' And she performed an Act of Truth. The instant she performed
her Act of Truth that mighty Ganges flowed back upstream with a roar,
in the sight of all that mighty throng.

"When the king heard the roar caused by the movement of the whirl-
pools and the waves of the mighty Ganges, he was astonished, and filled
with wonder and amazement. Said he to his ministers, 'How comes it
that this mighty Ganges is flowing back upstream?' 'Your Majesty, the
courtesan Bindumati heard your words, and performed an Act of Truth.
It is because of her Act of Truth that the mighty Ganges is flowing
backward.'

* From Heinrich Zimmer, *Philosophies of India,* ed. by Joseph Campbell
(New York: Bollingen Foundation, 1951), pp. 161–169. Copyright 1951 by
Bollingen Foundation; distributed by Pantheon Books.

"His heart palpitating with excitement, the king himself went posthaste and asked the courtesan, 'Is it true, as they say, that you, by an Act of Truth, have made this river Ganges flow back upstream?' 'Yes, Your Majesty.' Said the king, 'You have power to do such a thing as this! Who, indeed, unless he were stark mad, would pay any attention to what you say? By what power have you caused this mighty Ganges to flow back upstream?' Said the courtesan, 'By the Power of Truth, Your Majesty, have I caused this mighty Ganges to flow back upstream.'

"Said the king, 'You possess the Power of Truth! You, a thief, a cheat, corrupt, cleft in twain, vicious, a wicked old sinner who have broken the bounds of morality and live on the plunder of fools!' 'It is true, Your Majesty; I am what you say. But even I, wicked woman that I am, possess an Act of Truth by means of which, should I so desire, I could turn the world of men and the worlds of the gods upside down.' Said the king, 'But what is this Act of Truth? Pray enlighten me.'

" 'Your Majesty, whosoever gives me money, be he a Ksatriya or a Brahman or a Vaisya or a Sudra, or of any other caste soever, I treat them all exactly alike. If he be a Ksatriya, I make no distinction in his favor. If he be a Sudra, I despise him not. Free alike from fawning and contempt, I serve the owner of the money. This, Your Majesty, is the Act of Truth by which I caused the mighty Ganges to flow back upstream.' "

Just as day and night alternate, each maintaining its own form, and support by their opposition the character of the processes of time, so in the sphere of the social order everyone sustains the totality by adhering to his own dharma. The sun in India withers vegetation, but the moon restores it, sending the revivifying dew; similarly, throughout the universe the numerous mutually antagonistic elements cooperate by working against each other. The rules of the castes and professions are regarded as reflections in the human sphere of the laws of this natural order; hence, when adhering to those rules the various classes are felt to be collaborating, even when apparently in conflict. Each race or estate following its proper righteousness, all together do the work of the cosmos. This is the service by which the individual is lifted beyond the limitations of his personal idiosyncrasies and converted into a living conduit of cosmic force.

The Sanskrit noun *dharma*, from the root *dhr*, "to hold, to bear, to carry" (Latin *fero*; cf. Anglo-Saxon *faran*, "to travel, to fare"; cf. also, "ferry"), means "that which holds together, supports, upholds." Dharma refers not only to the whole context of law and custom (religion, usage, statute, caste or sect observance, manner, mode of behavior, duty, ethics, good works, virtue, religious or moral merit, justice, piety, impartiality), but also to the essential nature, character, or quality of the individual, as a result of which his duty, social function, vocation, or moral standard is what it is. Dharma is to fail just before the end of the world, but will endure as long as the universe endures; and each participates in its power as long as he plays his role. The word implies not only a universal law by which the cosmos is governed and sustained, but also particular laws, or inflections of "the law," which are natural to each special species or modification of existence. Hierarchy, specialization, one-sidedness, traditional obligations, are thus of the essence of the system. But there is no class struggle; for one cannot strive to be something other than what one is. One either "is" (*sat*) or one "is not" (*a-sat*), and one's dharma is the form of the manifestation in time of what one *is*. Dharma is ideal justice made alive; any man or thing without its dharma is an inconsistency. There are clean and unclean professions, but all participate in the Holy Power. Hence "virtue" is commensurate with perfection in one's given role.

The turbaned queen—so runs another tale—longing to greet the sage, her husband's brother, bade farewell to the king, her husband, and at eventide took the following vow: "At early morn, accompanied by my retinue, I will greet the sage Soma and provide him with food and drink; only then will I eat."

But between the city and the forest there was a river; and in the night there was a freshet; and the river rose and swept along, both strong and deep. Disturbed by this, when morning came, the queen asked her beloved husband, "How can I fulfill this my desire today?"

Said the king, "O queen, be not thus distressed, for this is simple to do. Go, easy in mind, with your retinue, to the hither bank; and, standing there, first invoke the goddess of the river, and then, with hands both joined, and with a pure heart, utter these words: 'O river goddess, if from the day my husband's brother took his vow, my husband has lived chaste, then straightway give me passage.'"

Hearing this, the queen was astonished, and thought, "What manner of thing is this? The king speaks incoherently. That from the day of his brother's vow the king has begotten progeny of sons on me, all this signifies that I have performed to him my vow as a wife. But after all, why doubt? Is physical contact in this case the meaning intended? Besides, women who are loyal to their husbands should not doubt their husbands' words. For it is said: A wife who hesitates to obey her husband's command, a soldier who hesitates at his king's command, a pupil who hesitates at his father's command, such a one breaks his own vow."

Pleased at this thought, the queen, accompanied by her retinue in ceremonial attire, went to the bank of the river, and standing on the shore did worship, and with a pure heart uttered distinctly the proclamation of truth recited by her husband.

And of a sudden the river, tossing its waters to the left and to the right, became shallow and gave passage. The queen went to the farther shore, and there, bowing before the sage according to form, received his blessing, deeming herself a happy woman. The sage then asked the woman how she had been able to cross the river, and she related the whole story. Having so done, she asked the prince of sages, "How can it be possible, how can it be imagined, that my husband lives chaste?"

The sage replied, "Hear me, good woman. From the moment when I took my vow, the king's soul was free from attachment and vehemently did he long to take a vow. For no such man as he could patiently endure to bear the yoke of sovereignty. Therefore he bears sway from a sense of duty, but his heart is not in what he does. Moreover it is said, 'A woman who loves another man follows her husband. So also a yogi attached to the essence of things remains with the round of existences.' Precisely so the chastity of the king is possible, even though he is living the life of a householder, because his heart is free from sin, just as the purity of the lotus is not stained, even though it grow in the mud."

The queen bowed before the sage, and then, experiencing supreme satisfaction, went to a certain place in the forest and set up her abode. Having caused a meal to be prepared for her retinue, she provided food and drink for the sage. Then, her vow fulfilled, she herself ate and drank.

When the queen went to take leave of the sage she asked him once more, "How can I cross the river now?" The sage replied, "Woman of tranquil speech, you must thus address the goddess of the river: 'If this sage, even to the end of his vow, shall always abide fasting, then grant me passage.'"

Amazed once more, the queen went to the bank of the river, proclaimed the words of the sage, crossed the river, and went home. After relating the whole story to the king, she asked him, "How can the sage be fasting, when I myself caused him to break the fast?"

The king said, "O queen, you are confused in mind; you do not understand in what true religion consists. Tranquil in heart, noble in soul is he, whether in eating or in fasting. Therefore: even though a sage eat, for the sake of religion, food which is pure, which he has neither himself prepared, nor caused another to prepare, such eating is called the fruit of a perpetual fast. Thought is the root, words are the trunk, deeds are the spreading branches of religion's tree. Let its roots be strong and firm, and the whole tree will bear fruit."

The visible forms of the bodies that are the vehicles of the manifestation of dharma come and go; they are like the falling drops of rain, which, ever passing, bring into sight and support the presence of the rainbow. What "is" (*sat*) is that radiance of being which shines through the man or woman enacting perfectly the part of dharma. What "is not"

42

(*a-sat*) is that which once was not and soon will not be; namely, the mere phenomenon that seems to the organs of sense to be an independent body, and therewith disturbs our repose by arousing reactions—of fear, desire, pity, jealousy, pride, submission, or aggression—reactions addressed, not to what is made manifest, but to its vehicle. The Sanskrit *sat* is the present participle of the verbal root *as*, "to be, to exist, to live"; *as* means "to belong to, to be in the possession of, to fall to the share of"; also "to happen to or to befall anyone, to arise, spring out, occur"; *as* means "to suffice," also "to tend to, to turn out or prove to be; to stay, reside, dwell; to be in a particular relation, to be affected." Therefore *sat*, the present participle, means, literally, "being, existing, existent"; also "true, essential, real." With reference to human beings, *sat* means "good, virtuous, chaste, noble, worthy; venerable, respectable; learned, wise." *Sat* means also "right, proper, best, and excellent," as well as "handsome, beautiful." Employed as a masculine noun, it denotes "a good or virtuous man, a sage"; as a neuter noun, "that which really exists, entity, existence, essence; reality, the really existent truth; the Good"; and "Brahman, the Holy Power, the Supreme Self." The feminine form of the noun, *sati*, means "a good and virtuous wife" and "a female ascetic." Sati was the name assumed by the universal Goddess when she became incarnate as the daughter of the old divinity Daksa in order to become the perfect wife of Siva. And *sati*, furthermore, is the Sanskrit original form of the word that in English now is "suttee," denoting the self-immolation of the Hindu widow on her husband's funeral pyre—an act consummating the perfect identification of the individual with her role, as a living image of the romantic Hindu ideal of the wife. She is the goddess Sati herself, reincarnate; the *sakti*, or projected life-energy, of her spouse. Her lord, her enlivening principle, having passed away, her remaining body can be only *a-sat*, non-*sat*: "unreal, non-existence, false, untrue, improper; not answering its purpose; bad, wicked, evil, vile." *Asat*, as a noun, means "nonexistence, nonentity; untruth, falsehood; an evil," and in its feminine form, *asati*, "an unchaste wife."

The tale of the queen, the saint, and the king teaches that Truth (*sat-ya*: "is-ness") must be rooted in the heart. The Act of Truth has to build out from there. And consequently, though dharma, the fulfillment of one's inherited role in life, is the traditional basis of this Hindu feat of virtue, nevertheless, a heart-felt truth of any order has its force. Even a shameful truth is better than a decent falsehood—as we shall learn from the following witty Buddhist tale.

The youth Yannadatta had been bitten by a poisonous snake. His parents carried him to the feet of an ascetic, laid him down, and said, "Reverend sir, monks know simples and charms; heal our son."

"I know no simples; I am not a physician."

"But you are a monk; therefore out of charity for this youth perform an Act of Truth."

The ascetic replied, "Very well, I will perform an Act of Truth." He laid his hand on Yannadatta's head and recited the following stanza:

> For but a week I lived the holy life
> With tranquil heart in quest of merit.
>
> The life I've lived for fifty years
> Since then, I've lived against my will.
>
> By this truth, health!
> Poison is struck down! Let Yannadatta live!

Immediately the poison came out of Yannadatta's breast and sank into the ground.

The father then laid his hand on Yannadatta's breast and recited the following stanza:

> Never did I like to see a stranger
> Come to stay. I never cared to give.
>
> But my dislike, the monks and Brahmans
> Never knew, all learned as they were.
>
> By this truth, health!
> Poison is struck down! Let Yannadatta live!

Immediately the poison came out of the small of Yannadatta's back and sank into the ground.

The father bade the mother perform an Act of Truth, but the mother replied, "I have a Truth, but I cannot recite it in your presence."

The father answered, "Make my son whole anyhow!" So the mother recited the following stanza:

> No more, my son, do I now hate this snake malignant
> That out of a crevice came and bit you, than I do your father!
>
> By this truth, health!
> Poison is struck down! Let Yannadatta live!

Immediately the rest of the poison sank into the ground, and Yannadatta got up and began to frisk about.

This is a tale that could be taken as a text for psychoanalysis. The opening up of the repressed truth, deeply hidden beneath the years of lies and dead actions that have killed the son (i.e., have killed the future, the life, of this miserable, hypocritical, self-deceiving household), suffices, like magic, to clear the venom from the poor, paralyzed body, and then all of that deadness (*asat*), "nonexistence," is truly nonexistent. Life breaks forth anew, in strength, and the living is spliced back to what was living. The night of nonentity between is gone.

❀

## THE RIG-VEDA

The people of the Rig-Veda were a rather light-skinned people who spoke an Aryan, or Indo-European, tongue, and worshiped the sun, the vault of heaven, fire, soma (the sacrificial drink), and a number of other personalized forces. As a practical, martial people, they conquered and organized a vast area and kept pushing on south and east to consolidate their physical and spiritual domain.

Three of the hymns of the Rig-Veda, one to the god Savitar, one to the god Indra, one to the spirit of man, show us the exuberant tone of their life.

### To Savitar*

The god his mighty hand, his arm outstretches
In heaven above, and all things here obey him;
To his commands the waters are attentive,
And even the rushing wind subsides before him.

Driving his steeds, now he removes the harness,
And bids the wanderer rest him from his journey.
He checks the serpent-smiter's eager onset;
At Savitar's command the kindly Night comes.

The weaver rolls her growing web together,
And in the midst the workman leaves his labor;
The god arises and divides the seasons,
God Savitar appears, the never resting.

In every place where mortals have their dwelling,
The house-fire far and wide sheds forth its radiance.
The mother gives her son the fairest portion,
Because the god has given desire of eating.

* From Adolf Kaegi, *The Rigveda: The Oldest Literature of the Indians*, trans. by R. Arrowsmith (Boston: Ginn and Company, 1886).

Now he returns who had gone forth for profit;
For home the longing wanderer's heart is yearning;
And each, his task half finished, homeward journeys.
This is the heavenly Inciter's ordinance.

The restless, darting fish, at fall of evening,
Seeks where he may his refuge in the waters,
His nest the egg-born seeks, their stall the cattle;
Each in his place, the god divides the creatures.

### To Indra*

He who, just born, chief god of lofty spirit by power and
     might became the gods' protector,
Before whose breath through greatness of his valor the two
     worlds trembled, he, O men, is Indra.

He who fixed fast and firm the earth that staggered, and set
     at rest the agitated mountains,
Who measured out the air's wide middle region and gave the
     heaven support, he, men, is Indra.

Who slew the dragon, freed the seven rivers, and drove the
     kine forth from the cave of Vala,
Begat the fire between two stones, the spoiler in warrior's
     battle, he, O men, is Indra.

By whom this universe was made to tremble, who chased
     away the humbled brood of demons,
Who, like a gambler gathering his winnings, seized the foe's
     riches, he, O men, is Indra.

Of whom, the terrible, they ask, Where is he? or verily they
     say of him, He is not.
He sweeps away, like birds, the foe's possessions. Have faith
     in him, for he, O men, is Indra.

Stirrer to action of the poor and lowly, of priest, of suppliant
     who sings his praises;
Who, fair-faced, favors him who presses soma with stones
     made ready, he, O men, is Indra.

* "To India" and the following "Hymn of Man," trans. by Ralph J. Griffith,
are from Lin Yutang (ed.), *The Wisdom of India and China* (New York:
Random House, 1942). Copyright 1942 by Random House, Inc. Reprinted by
permission.

He under whose supreme control are horses, all chariots, and
    the villages, and cattle;
He who gave being to the sun and morning, who leads the
    waters, he, O men, is Indra.

To whom two armies cry in close encounter, both enemies
    the stronger and the weaker;
Whom two invoke upon one chariot mounted, each for him-
    self, he, O ye men, is Indra.

Without whose help our people never conquer; whom, bat-
    tling, they invoke to give them succor;
He of whom all this world is but the copy, who shakes things
    moveless, he, O men, is Indra.

He who hath smitten, ere they knew their danger, with his
    hurled weapon many grievous sinners;
Who pardons not his boldness who provokes him, who slays
    the Dasyu, he, O men, is Indra.

He who discovered in the fortieth autumn Sambara as he
    dwelt among the mountains;
Who slew the dragon putting forth his vigor, the demon lying
    there, he, men, is Indra.

Who with seven guiding reins, the bull, the mighty, set free
    the seven great floods to flow at pleasure;
Who, thunder-armed, rent Rauhina[1] in pieces when scaling
    heaven, he, O ye men, is Indra.

Even the heaven and earth bow down before him, before his
    very breath the mountains tremble.
Known as the soma-drinker, armed with thunder, who wields
    the bolt, he, O ye men, is Indra.

Who aids with favor him who pours the soma and him who
    brews it, sacrificer, singer,
Whom prayer exalts, and pouring forth of soma, and this our
    gift, he, O ye men, is Indra.

Thou verily art fierce and true who sendest strength to the
    man who brews and pours libation.
So may we evermore, thy friends, O Indra, speak loudly to
    the synod with our heroes.

[1] A demon of drought.

# THE PSYCHOLOGY OF INDIA

### Hymn of Man

A thousand heads hath Purusha,[2] a thousand eyes, a thousand feet.
On every side pervading earth he fills a space ten fingers wide.

This Purusha is all that yet hath been and all that is to be,
The lord of immortality which waxes greater still by food.

So mighty is his greatness; yea, greater than this is Purusha.
All creatures are one-fourth of him, three-fourths eternal life in heaven.

With three-fourths Purusha went up: one-fourth of him again was here.
Thence he strode out to every side over what eats not and what eats.

From him Viraj[3] was born; again Purusha from Viraj was born.
As soon as he was born he spread eastward and westward o'er the earth.

When gods prepared the sacrifice with Purusha as their offering,
Its oil was spring; the holy gift was autumn; summer was the wood.

They balmed as victim on the grass Purusha born in earliest time.
With him the deities and all Sadhyas[4] and Rishis sacrificed.

From that great general sacrifice the dripping fat was gathered up.
He formed the creatures of the air, and animals both wild and tame.

From that great general sacrifice Richas and Sama-hymns were born:
Therefrom were spells and charms produced; the Yajus had its birth
  from it.

From it were horses born, from it all cattle with two rows of teeth:
From it were generated kine, from it the goats and sheep were born.

When they divided Purusha, how many portions did they make?
What do they call his mouth, his arms? What do they call his thighs and
  feet?

The Brahman[5] was his mouth, of both his arms was the Rajanya[6] made.
His thighs became the Vaisya,[7] from his feet the Sudra[8] was produced.

---

[2] Embodied spirit, or man personified and regarded as the soul and original source of the universe, the personal and life-giving principle in all animated beings.
[3] One of the sources of existence.
[4] Celestial beings.
[5] The first caste of Brahman priests.
[6] The second caste of kings.
[7] The third caste of traders.
[8] The fourth caste of laborers.

The moon was gendered from his mind, and from his eye the sun had
    birth;
Indra and Agni from his mouth were born, and Vayu[9] from his breath.

Forth from his navel came mid-air; the sky was fashioned from his head;
Earth from his feet, and from his ear the regions. Thus they formed the
    worlds.

Seven fencing-sticks had he, thrice seven layers of fuel were prepared
When the gods, offering sacrifice, bound, as their victim, Purusha.

Gods, sacrificing, sacrificed the victim: these were the earliest holy
    ordinances.
The mighty ones attained the height of heaven, there where the Sadhyas,
    gods of old, are dwelling.

The people of the Rig-Veda thought of themselves as superior.
Until the discoveries (see page 5) in the western part of the Indian
subcontinent, Europeans had often believed the Aryans were
correct in their estimation of themselves; that the Aryans were, in
fact, superior to the dark-skinned people whom they overran. It
has become clear, however, from the diggings and reconstruc-
tions that Indian civilization was already rich, complex, and
fertile in philosophical ideas long before the people of the Rig-
Veda arrived on the scene. Certainly, however, interaction occurred
and certainly the worship of the forces of nature continued to express many of the ideas of the Vedic peoples. It took several centuries for the robust, self-assertive philosophy of the Aryan-speaking
peoples to become infiltrated with the more serious or even solemn
—or pessimistic—philosophical spirit which prevailed in the Indian
peninsula. At any rate, the Sanskrit language which voices, through
exuberant hymns, the self-assurance of the conqueror in the new
land begins within a few centuries to express the frustrations,
doubts, and perplexities of thoughtful men; and the solution for
life's difficulties and contradictions is sought in reflection, in self-
discipline, and in forms of inward turning which one can only call
psychological.*

    [9] God of Wind.
    * Cf. Heinrich Zimmer, *Philosophies of India* (New York: Bollingen Founda-
tion, 1951), pp. 333–355.

In the West we ordinarily think of Indian philosophy as "ideal-istic," or as emphasizing the reality of the soul and the changeless character of the absolute. It is, however, of interest to anyone dealing with "the meeting of East and West" to note that the Indians, no less than the Greeks, developed a wide range of philosophical ideas, representing in fact many of the same basic conceptions that have developed from Greek philosophy through modern times in the European tradition. The peoples who composed the Vedic hymns, the people of the Rig-Veda who conquered the highly civilized native populations, both gave and received philo-sophical ideas in interaction with them; emphasized the seriousness, the solemnity, the unchangeability, the timelessness of man's spiritual nature; and, as we shall see, defined the many paths by which man may find unity with an eternal spiritual principle. It is just as important, however, to note that Indian philosophy dis-covered materialism, mechanism, the pleasure-pain philosophy ("hedonism"); complex things were reduced to simple, and simple were often made very non-spiritual indeed.

Side by side, moreover, with these mechanistic or hedonistic philosophies, there developed a very rich practical philosophy of everyday living, expressed in the cynical or comic language of the great storybook The Panchatantra; and practical advice to the king, the courtier, and the plain man is to be found almost every-where. The art of lovemaking is spelled out with extraordinary audacity, and at times much beauty and much "common sense" combined. We have used just enough of this "non-spiritual," "non-idealistic" to give flavor.

But we must give the Upanishads their central place. For what is really distinctive and different about Indian philosophy is its unique definition of the *spiritual*, and we shall therefore use our limited space primarily for this purpose, while reminding the reader that just as Hebrews, Greeks, Romans, early Christians, and medi-eval peoples all had many facets and psychologized in many differ-ent ways, so the Indians psychologized in many different ways. Indeed, as we go on to China and Japan, we shall find likewise that we must select major distinctive trends, but not forget that there is a wide diversity in any and all of the great national psychologies.

## THE UPANISHADS

The great philosophical treatises known as the Upanishads, which began to take shape around 800 B.C. and developed for several hundred years, form the heart of Indian psychology. They were composed after the period of the exuberant Vedic hymns, as life became stabilized in the Indian peninsula and as a warrior caste gave itself more and more to contemplation and to cultivation of the inner life.

The following is an interpretative introduction to the Upanishads and their psychological aspects by our Indian consultant, Professor B. Kuppuswamy of the India International Centre, New Delhi.

The Upanishads form the fourth part of the Vedas. The word "Upanishad" means, literally, "secret teaching." Possibly they were secret instructions to persons who were eager to learn and who were also mentally and morally worthy of the teaching. There are many Upanishads, but only about a dozen are regarded as classic. The principal Upanishads have a wealth of material that is of psychological interest. As an example of the texts and their interpretation, I will begin with the Katha Upanishad, in an attempt to bring together the various passages in the Katha Upanishad that are of value to modern students of psychology and philosophy. In doing so it is inevitable that some kind of systematization is made where none was intended by the author of the Upanishad.

### The Katha Upanishad

THE SELF, MIND, AND SENSES

The analogy of the chariot (I.3) shows the interrelationship between the self (*atman*), the intellect (*buddhi*), the mind (*manas*), and the senses (*indriyas*).

> Know the Self as the lord of the chariot and the body as, verily, the chariot, know the intellect as the charioteer and the mind as, verily, the reins.

51

The senses, they say, are the horses; the objects of sense the paths (they range over); (the self) associated with the body, the senses, and the mind—wise men declare—is the enjoyer.*

These two passages set down the analysis of the various aspects of mind. For the Upanishadic thinkers the senses, the *indriyas*, include both the cognitive and the executive organs: what are called the *gnanendriyas, karmendriyas*: speech, manipulation, walking, evacuation, and reproduction (the generative organs). Another distinctive feature of these two passages is the differentiation between the *manas*, or mind; the *buddhi*, or discriminative intellect; and the self, or *atman*. Yet another feature is that the ultimate development of the human body consists in developing the self as the sole controller, using discrimination and the mind in controlling the cognitive as well as the executive sense organs, which are ever ready to pursue their own ends.

In the next two passages the analogy of the chariot is extended to show that in an individual who has no understanding (*avignanavan*) and whose *manas* is unrestrained, and his senses out of control, the senses behave like wild horses, while the person with understanding and whose *manas* is restrained will have all his senses under his control. The individual who has understanding and the control of his sense organs reaches the end of his journey, the goal of human development.

A hierarchy is established to show that the objects of the senses are higher than the senses themselves; that *manas* is higher than the objects that the senses seek; that the *buddhi* (discrimination) is higher than the *manas*; and that the great self is higher than the *buddhi*. The wise man is one who restrains his speech in the mind, restrains the mind in the understanding, and the understanding in the self. This leads to *shanti*, tranquillity and peace. A man reaches the highest state (*paramangatim*) when the five cognitive sense organs and the *manas* cease from their normal activities and the *buddhi* does not stir.

### THE DESIRES AND THE PLEASURE PRINCIPLE

When Nachiketas requests Yama to teach him the highest knowledge, Yama tries to avoid this by offering a number of temptations. He asks Nachiketas to choose vast expanses of land, cattle and horses and elephants, gold, sons and grandsons who will live for a long time, long life for himself, and noble maidens with chariots and musical instruments: "I shall make thee the enjoyer of thy desires." However,

* All quotations from the Upanishads in this volume are from S. Radhakrishnan, *The Principal Upanishads* (London: George Allen & Unwin; New York: Harper & Row, 1953). Copyright 1953 by Harper & Brothers.

Nachiketas is determined to attain knowledge, and refers to the fact that life itself is brief and all pleasure, arising out of the senses, is transitory. By renouncing pleasure Nachiketas has avoided the path which generally leads human beings to ruin. The man who is thoughtful and wise ponders over the good as well as the pleasant, and chooses the good in preference to the pleasant, whereas the simple-minded, the dull, and the stupid person will choose the pleasant.

## SELF-REALIZATION

There are many passages in the Katha Upanishad which set forth the characteristics of a man who has achieved self-realization.

In I.1.10 and 11, Nachiketas requests the boon that will enable his father to be tranquil in his thoughts and serene in mind and to be *free from passions*. With freedom from anger he should be able to sleep peacefully through the night. When a man strips himself of the active will (*a-kratu*), he can uphold his self, in freedom from sorrow. "When sitting he moves far; when lying he goes everywhere." In other words, he transcends the limitations imposed by time and space.

## THE MEANS TO SELF-REALIZATION

Self-realization can be achieved only through mind (*manasa veva*). The self cannot be attained by instruction (*pravachana*), or by intellectual power (*medhasakti*), or by hearing (*srutena*). The self can be attained only through the self, by the choice of the self (*atman vivrunute tanun svaam*).

There is an apparent contradiction, according to Indian terminology, between II.1.11, which asserts that self-realization can be achieved only through the *manas*, or mind, and I.2.23, which asserts that it can be attained only through the *atman*, or self. There is also another apparent contradiction, in that self-realization occurs spontaneously and does not arise either from instruction or from intellect.

In five passages (in the third *valli* of the second chapter) a brief description of yoga is given. In II.3.10 it is asserted that, when the five cognitive senses and the *manas* and the *buddhi* cease from their normal activities, the individual reaches the highest state (*paramangatim*). This state of steady control of the senses is what is called yoga. It is reiterated that the self cannot be realized through either speech, mind, or sight. It can only be affirmed that "he is" (*asti iti*). It is also stated that he should be apprehended only as existent. This state is possible when all the desires (*kamas*) are cast away, when all the complexes (*hridayasyeha grantayaha*) are rent asunder (*pravidhyante*).

53

The following are representative passages (II.1.1 and II.3.7):

> The Self is not to be sought through the senses. The Self-caused pierced the openings (of the senses) outward; therefore one looks outward and not within oneself. Some wise man, however, seeking life eternal, with his eyes turned inward, saw the Self.

> Beyond the senses is the mind; above the mind is its essence (intelligence); beyond the intelligence is the great Self; beyond the great (Self) is the unmanifest.

The Katha Upanishad, then, discusses the relationship between the self, *atman*; the intellect, *buddhi*; the mind, *manas*; and the sense organs, the *indriyas*. It describes the different motives and desires which impel an individual to act. Such desires are classified into two groups, the good, the *shreyas*, and the pleasant, the *preyas*. The individual who chooses the path of the pleasant, the *preyas*, fails in his aim.

### Summary of the Principal Upanishads

INTERACTION OF MIND AND BODY

The Taittiriya Upanishad develops the doctrine of the *kosas*. Man is looked upon as being formed of five sheaths; the outermost is the physical (*annamaya*); within it is the biological (*pranamaya*); within that is the psychological (*manomaya*); within that is the intellectual (*virgnanamaya*); the innermost part is the sheath of bliss (*anandamaya*). This is an important contribution, for it shows that the highest level of personality can occur only in a person who is intellectually well developed, and that this development is conditioned by the physical, the biological, and the lower psychological elements. Thus this Upanishad gives an organismic view of behavior which involves the biological and physical.

Some suggestions that may be looked upon as showing the close interaction between mind and body can be found in the Mundaka Upanishad, which states that from food life arises and from life the mind.

Similarly the Maitri Upanishad shows the importance of food for psychological well-being. The Chandogya Upanishad speaks of the relationship between food and memory.

THE FOUR STATES

The Mundaka Upanishad asserts that the self has four quarters, namely, the waking state, the dream state, the deep-sleep state, and finally the state of self-realization. The Prasna Upanishad discusses the relationship between the waking state and the sleeping state. The Katha Upanishad asserts that a man can sleep peacefully through the night only when his anger has gone. The Kaushitaki Upanishad shows the relationships between the waking, dream, and sleeping states. It also (IV.20) states that when a man sleeps speech, hearing, sight, and thought are withdrawn and when he awakes they all reappear. In this same Upanishad (IV.19) we find that Ajatasatru demonstrates the condition of an individual who is asleep and shows how he can be made to wake up and understand who he is and what he is. There is a similar reference to the sleeping state in the Maitri Upanishad, the Brihad-Aranyaka Upanishad, and the Chandogya Upanishad.

THE SENSES, THE MIND, THE INTELLECT, AND THE SELF

The Brihad-Aranyaka Upanishad tells the story of the struggle between the gods and the demons for mastery of the world. In this story the gods manage to obtain speech, smell, sight, hearing, mind, and the self. The demons then succeed in piercing these faculties one after the other and create evil speech, evil smell, etc., but they can do nothing to the self. Further on, the same Upanishad shows that the mind is responsible for perception; mere stimulation of the sense organs is not enough. It also gives the names of the various mental activities, such as desire, determination, doubt, faith, steadfastness, shame, intellection, and fear. It addresses itself to specific psychological questions: because of the mind one can know when one is touched on his back.

Another passage shows how speech and the various sense organs were produced. As these sense organs become weary, death takes possession of them, but the self is something which never becomes weary and so is deathless. A later passage speaks of the various sense organs as well as the organs of action and of the mind.

This ancient Upanishad classifies the *indriyas* into the sensory and executive. Another passage speaks of the self as a person "who consists of the knowledge among senses" and also as one who moves between two worlds, between waking and sleeping. It also speaks of the dream state as an intermediate third state. It refers to the way in which the self is able to create objects as well as feelings in the dream state, and asserts that sex and fear are the dominant features of the dream state.

The Prasna Upanishad also speaks of the relationship between the various physical aspects, the sense organs and the organs of action. It attributes perception to the mind, conception to the intellect, and illumination to the self. Both the Kaushitaki Upanishad and the Maitri Upanishad refer to the relationship between organs of sense and organs of action, the mind, the intellect, and the self.

The Chandogya Upanishad looks upon the problem in a slightly different way. The universe has four quarters: the speech, the breath, the eye, and the ear. It refers to the state of sleep as something which absorbs speech, sight, hearing, and mind into the breath.

There is an interesting story in this Upanishad about a dispute among the various sense organs as to their respective importance. They go to their father and ask him to settle the dispute. He says that that sense organ is the best whose departure makes the body look the worst. Speech departs for a year and finds on return that there is no difference in the individual. Next the eye departs and then the ear and then the mind. But when the breath tried to depart, all the other senses realized that the breath is the most important. The interesting part of the story is the statement that when the mind departs the individual lives like a child, "mindless but breathing with breath, speaking with speech, seeing with the eye, hearing with the ear." Thus this passage gives an inkling of the realization that even when the mental activities are not operating the other sense organs may be functioning, though at a lower level.

### PERCEPTION AND INTERPRETATION

The Brihad-Aranyaka Upanishad reveals that the Upanishadic thinkers had some insight into the influence of individuality on perception. The gods, men, and the demons are said to have gone to their father to learn. At the end of their course of instruction, the teacher uttered the syllable *da* and asked each group whether they had understood. The gods answered that he had taught them *damyata*, meaning control over self. The men understood it as giving charity, and the demons in their turn said they understood it as *dayadhvam*—to be compassionate. Thus each group interpreted the syllable in terms of its own lack, namely, self-control, charity, and compassion.

### THE SELF

The Aitareya Upanishad asks the question "If each action arises through an organ of action and if each sensation arises through an organ of sense, who am I?" The reply is given in another passage, where it is said that the self is the one who sees, who speaks, who discriminates, etc. A similar question is asked in the Kena Upanishad. The self is described as the ear of the ear, the mind of the mind, and so on. It is also asserted that because the eye cannot see the self, or the speech speak about it, or the mind or the intellect understand it, it is very difficult to teach anything about it; so it is said that the self is other than the known as well as the unknown. In another passage the Upanishad tries to teach concerning the self. It is looked upon as an agency for memory; and volition, austerity, self-control, and work are its support.

The Mundaka Upanishad speaks of the self as composed of two aspects, using an analogy of two birds: they live on the same tree, one eating the sweet fruit, and the other looking on. In the next passage it is asserted that the person who is immersed in life is deluded, while the person who sees himself as participating in the activities of life is free from sorrow. The Svetasvatara Upanishad also uses the analogy of two birds, and employs almost the same words as the Mundaka Upanishad. This analogy shows that the Upanishadic thinkers were aware of two aspects of the self, one which is engaged in day-to-day routine activities and the other which is a detached observer. Our trouble is that we identify ourselves only with the former aspect of the self and so become involved in our joys and sorrows, pains and pleasures, and failures and successes.

The Kaushitaki Upanishad stresses that one should try to understand not the speech or the sensation but the person who speaks and the person who lives through the sensation. Similarly we must understand not the deed but the doer, and so on. It also refers to all these various aspects as spokes that are fixed to the self. The Maitri Upanishad gives a long description of the relationship between the physical elements of the body and the self, and depicts the self as like a drop of water on a lotus leaf— on it, but not attached to it. It also proceeds to explain that our confusion and bewilderment arise from a failure to realize this relationship; self-love is responsible for this. "Thinking, 'I am he,' 'This is mine,' he binds himself with his self like a bird in a snare."

The Brihad-Aranyaka Upanishad contains many passages that refer to the notion of the self. It asserts that self-awareness—the sense of I— is the basis for one's idea of the self, but self-awareness leads to fear arising out of a sense of loneliness; then the various aspects of breath, speech, mind, etc., appear. There is a long list of mental activities— desire, determination, doubt and faith, etc. One of the most famous passages of this Upanishad describes how everything is looked upon as dear because of one's love for one's self. Consequently, one should meditate upon and understand the self. It also indicates that it is the sense of duality that is responsible for various aspects. The self is the inner controller of all experiences. Self is that which transcends hunger, thirst, sorrow and delusion, old age and death. In another passage it states that one becomes what one knows and understands. The Chandogya Upanishad refers to the self as something that consists of the mind, body, etc. It also refers to the self as something that is free from desire, grief, old age, and death. It tells a story in which the gods and the demons try to find the self. The teacher asks them to look at the reflection in the water or a mirror or an eye and know the self. The demons appear to be quite satisfied with this explanation, but the gods are not. The teacher then asks them to recognize the self in the experience of their dreams, but even this is not satisfactory. They go to the teacher again and learn that the self is neither the body nor a free person who dreams, but that on which is based all the experiences of the waking as well as the dreaming state.

One striking feature of Upanishadic thought is the characterization of the self with contradictory attributes. The Mundaka Upanishad says it is "farther than far, yet here near at hand." It also asserts that the self cannot be grasped by speech or by any sense organs; nor can it be realized by austerity or by work. It is only when one's intellect is developed, and by meditation, that one can grasp it. Similarly the Katha Upanishad speaks of it as "smaller than the small and greater than the great"; also, "sitting he moves far; lying he goes everywhere." The Svetasvatara Upanishad asserts that "without foot or hand (yet) swift and grasping, he sees without eye, he hears without ear. He knows whatever is to be known; of him there is none who knows." Moreover, the self is "subtler than the subtle, greater than the great." Similarly, the Chandogya Upanishad says that the self is smaller than the smallest grain and greater than the earth or the sky.

GOALS OF LIFE

One of the most significant contributions of Upanishadic thought is the distinction made in the Katha Upanishad with respect to the goals of life. Some men choose the path of pleasures (*preyas*), but wise men choose the path of good (*sreyas*). It also says that wise men choose good in preference to pleasure, whereas misguided persons prefer the pleasures of life. This Upanishad catalogues the pleasures of life—sex, progeny, land, cattle, gold, long life, music, chariots, and so forth—which people seek to enjoy. However, all these pleasures are transitory and so should not be sought after. The pleasures of life described in the Kaushitaki Upanishad are: sex, fruits, ointment, garland, garments, and perfumes. The Chandogya Upanishad also supplies a list of pleasures.

LIBERATION

The Katha Upanishad asserts that the self cannot be understood through the sense organs because sense organs can only help us to know what goes on in the external world. "Some wise man, however, seeking life eternal, with his eyes turned inward, saw the self." Thus the full development of the personality is possible only when the individual learns to look inward and becomes aware of his self. The same Upanishad also asserts, "This self cannot be attained by instruction . . ." This can be done only "when all the desires that dwell within the human heart are cast away . . . when all the knots that fetter here the heart are cut asunder"; through this process, when a person is freed from his passions, he is in a position to know himself.

The Mundaka Upanishad states that immature persons live in ignorance and perform rituals because of their *attachment*. On the other hand, those who practice austerity and become tranquil will attain liberation. Thus *non-attachment* is the important quality. It also asserts that the self-realized man will not talk of anything else. He performs his usual work but all the time he is "delighting in the self." The self can be realized by truth, right knowledge, austerity, and constant practice. Those who have attained the self become free from passion, and are tranquil. They are no longer distracted and can perform their work with concentration. According to the Isa Upanishad, such a man will feel no revulsion toward any thing because he "sees all beings in his own self and his own self in all beings." He can have neither delusions nor sorrow. The Taittiriya Upanishad claims that such a man will have no fear whatever, nor will he have any sense of guilt. The Kaushitaki Upanishad asserts that an individual, after he has satisfied hunger, sex, and other needs, proceeds to pay attention to eternal things; he shakes off his worldly attachment and transcends the pairs of opposites. The Svetasvatara Upanishad says that such an individual will control his mind and his thoughts. It also indicates the path to controlling posture and breath in order to attain liberation. Similarly, the Maitri Upanishad shows the value of the "control of the breath, withdrawal of the senses, meditation, concentration, contemplative inquiry, and absorption"; such a person will detach himself from good as well as evil.

The following extracts from several of the major Upanishads illustrate the psychological content of these classic treatises.

### MUNDAKA UPANISHAD

III.1.  3. When a seer sees the creator of golden hue, the Lord, the Person, the source of Brahma, then being a knower, shaking off good and evil and free from stain, he attains supreme equality with the lord.

4. Truly it is life that shines forth in all beings. Knowing him, the wise man does not talk of anything else. Sporting in the self, delighting in the self, performing works, such a one is the greatest of the knowers of *Brahman*.

5. This self within the body, of the nature of light and pure, is attainable by truth, by austerity, by right knowledge, by the constant (practice) of chastity. Him, the ascetics with their imperfections done away, behold.

6. Truth alone conquers, not untruth. By truth is laid out the path leading to the gods by which the sages who have their desires fulfilled travel to where is that supreme abode of truth.

7. Vast, divine, of unthinkable form, subtler than the subtle. It shines forth, farther than the far, yet here near at hand, set down in the secret place (of the heart) (as such) even here it is seen by the intelligent.

.         .         .

III.2.    2. He who entertains desires, thinking of them, is born (again) here and there on account of his desires. But of him who has his desire fully satisfied, who is a perfected soul, all his desires vanish even here (on earth).

3. This self cannot be attained by instruction nor by intellectual power nor even through much hearing. He is to be attained by the one whom (the self) chooses. To such a one the self reveals his own nature.

4. This self cannot be attained by one without strength nor through heedlessness nor through austerity without an aim. But he who strives by these means, if he is a knower, this self of his enters the abode of *Brahman*.

5. Having attained Him, the seers (who are) satisfied with their knowledge (who are) perfected souls, free from passion, tranquil, having attained the omnipresent (self) on all sides, those wise, with concentrated minds, enter into the All itself.

.         .         .

8. Just as the flowing rivers disappear in the ocean casting off name and shape, even so the knower, freed from name and shape, attains to the divine person, higher than the high.

## ISA UPANISHAD

I. 1. (Know that) all this, whatever moves in this moving world, is enveloped by God. Therefore find your enjoyment in renunciation; do not covet what belongs to others.

2. Always performing works here one should wish to live a hundred years. If you live thus as a man, there is no way other than this by which karman (or deed) does not adhere to you.*

. . .

6. And he who sees all beings in his own self and his own self in all beings, he does not feel any revulsion by reason of such a view.**

7. When, to one who knows, all beings have, verily, become one with his own self, then what delusion and what sorrow can be to him who has seen the oneness?

. . .

17. May this life enter into the immortal breath; then may this body end in ashes. O Intelligence, remember, remember what has been done. Remember, O Intelligence, what has been done. Remember.

## SVETASVATARA UPANISHAD

II. 8. Holding the body steady with the three (upper parts, chest, neck, and head) erect, causing the senses and the mind to enter into the heart, the wise man should cross by the boat of *Brahman* all the streams which cause fear.

9. Repressing his breathings here (in the body), let him who has controlled all movements, breathe through his nostrils, with diminished breath; let the wise man restrain his mind vigilantly as (he would) a chariot yoked with vicious horses.

10. In a level clean place, free from pebbles, fire and gravel, favorable to thought by the sound of water and other features, not offensive to the eye, in a hidden retreat protected from the wind, let him perform his exercises (let him practice yoga).

* That is, your acts will inevitably cling to you, bring consequences to you. [Eds.]

** "To be a brother to the insensible rock."—William Cullen Bryant, *Thanatopsis.*

11. Fog, smoke, sun, wind, fire, fireflies, lightning, crystal moon, these are the preliminary forms which produce the manifestation of *Brahman* in yoga.

. . .

13. Lightness, healthiness, steadiness, clearness of complexion, pleasantness of voice, sweetness of odor, and slight excretions, these, they say, are the first results of the progress of yoga.

14. Even as a mirror stained by dust shines brightly when it has been cleaned, so the embodied one when he has seen the (real) nature of the Self becomes integrated, of fulfilled purpose and freed from sorrow.

15. When by means of the (real) nature of his self he sees as by a lamp here the (real) nature of *Brahman*, by knowing God who is unborn, steadfast, free from all natures, he is released from all fetters.

. . .

III. 17. Reflecting the qualities of all the senses and yet devoid of all the senses, it is the lord and ruler, it is the great refuge of all.

18. The embodied soul in the city of nine gates sports (moving to and fro) in the outside (world), the controller of the whole world, of the stationary and the moving.

19. Without foot or hand, (yet) swift and grasping, he sees without eye, he hears without ear. He knows whatever is to be known; of him there is none who knows. They call him the Primeval, the Supreme Person.

20. Subtler than the subtle, greater than the great is the Self that is set in the cave of the (heart) of the creature. One beholds Him as being actionless and becomes freed from sorrow, when through the grace of the Creator he sees the Lord and His majesty.

. . .

V. 9. This living self is to be known as a part of the hundredth part of the point of a hair divided a hundredfold, yet it is capable of infinity.

10. It is not female, nor is it male; nor yet is this neuter. Whatever body it takes to itself, by that it is held.

11. By means of thought, touch, sight, and passions and by the abundance of food and drink there are the birth and development of the (embodied) self. According to his deeds, the embodied self assumes successively various forms in various conditions.

12. The embodied self, according to his own qualities, chooses (assumes) many shapes, gross and subtle. Having himself caused his union with them, through the qualities of his acts and through the qualities of his body, he is seen as another.

.      .      .

VI.   3. Having created this work and rested again, having entered into union with the essence of the self, by one, two, three, or eight, or by time too and the subtle qualities of the self.

.      .      .

19. To him who is without parts, without activity, tranquil, irreproachable, without blemish, the highest bridge to immortality like a fire with its fuel burned.

20. When men shall roll up space as if it were a piece of leather, then will there be an end of sorrow, apart from knowing God.

In our Judeo-Christian tradition, there have been periods of great joy, delight, and pride in the body; periods exemplified in such phrases as "the strong man delighting to run his race," "Who is she that looketh forth as the morning, fair as the moon, clear as the sun, and terrible as an army with banners?" The language of sculpture and painting is often the language of exultation in beauty and power. On the other hand, the belittlement of the body, the insistence that the body is of no interest or value, is equally striking, especially in the religious tradition; and often it goes beyond sheer indifference into some form of loathing or disgust. The repudiation of the body in terms of uncleanness, to be combined with evil, in such phrases as "the world, the flesh, and the devil," has led many to wonder if a good God could have created such unloveliness.

Now this curious duality or ambivalence toward the body appears likewise in the Indian tradition. The great epics of India, the Ramayana and the Mahabharata, are full of the lordly magnificence of godlike men and women, but the Upanishads and most of the other great philosophical works which follow this tradition are likewise full of belittlement, or even disgust, directed to the body. It is doubtful whether even the most intense Christian repudiation of the flesh could equal in intensity the following phrases from the Maitri Upanishad:

### MAITRI UPANISHAD

I.   3. O Revered One, in this foul-smelling, unsubstantial body, a conglomerate of bone, skin, muscle, marrow, flesh, semen, blood, mucus, tears, rheum, feces, urine, wind, bile, and phlegm, what is the good of the enjoyment of desires? In this body which is afflicted with desire, anger, covetousness, delusion, fear, despondency, envy, separation from what is desired, union with the undesired, hunger, thirst, old age, death, disease, sorrow, and the like, what is the good of the enjoyment of desires?

.    .    .

III. 2. There is, indeed, another, different, called the elemental self, he who, affected by the bright or the dark fruits of action, enters a good or an evil womb so that his course is downward or upward and he wanders about affected by the pairs (of opposites). And this is its explanation. The five subtle elements are called by the name element. Likewise the five gross elements are called by the name element. Now the combination of these is called the body. Now he, indeed, who is said to be in the body is called the elemental self. Now its immortal self is like a drop of water on the lotus leaf. This (elemental self) verily, is affected by nature's qualities. Now because of being affected, he gets to bewilderment (becomes confused); because of bewilderment he sees not the blessed Lord who dwells in himself, the causer of action. Borne along and defiled by the stream of qualities, unstable, wavering, bewildered, full of desire, distracted, he gets to the state of self-love. Thinking, "I am he," "This is mine," he binds himself with his self like a bird in a snare. So being affected by the fruits of his action, he enters a good or an evil womb so that his course is downward or upward and he wanders about, affected by the pairs of opposites.

## CHANDOGYA UPANISHAD

III.14. 1. Verily, this whole world is *Brahman*, from which he comes forth, without which he will be dissolved and in which he breathes. Tranquil, one should meditate on it. Now verily, a person consists of purpose. According to the purpose a person has in this world, so does he become on departing hence. So let him frame for himself a purpose.

2. He who consists of mind, whose body is life, whose form is light, whose conception is truth, whose soul is space, containing all works, containing all desires, containing all odors, containing all tastes, encompassing this whole world, being without speech and without concern.

3. This is my self within the heart, smaller than a grain of rice, than a barley corn, than a mustard seed, than a grain of millet, or than the kernel of a grain of a millet. This is myself within the heart, greater than the earth, greater than the atmosphere, greater than the sky, greater than these worlds.

4. Containing all works, containing all desires, containing all odors, containing all tastes, encompassing this whole world, without speech, without concern, this is the self of mine within the heart; this is *Brahman*.

. . .

V.1. 6. Now the (five) senses disputed among themselves as to who was superior, saying (in turn), "I am superior." "I am superior."

7. Those senses went to Praja-pati (their) father and said, "Venerable Sir, who is the best of us?" He said to them, "He on whose departing the body looks the worst, he is the best among you."

8. Speech departed and having stayed away for a year returned and said, "How have you been able to live without me?" (They replied) "Like the dumb not speaking, but breathing with the breath, seeing with the eye, hearing with the ear, thinking with the mind. Thus (we lived)." Speech entered in.

9. The eye departed and having stayed away for a year returned and said, "How have you been able to live without me?" (They replied) "Like the blind not seeing but breathing

65

with the breath, speaking with speech (the tongue), hearing with the ear, thinking with the mind. Thus (we lived)." The eye entered in.

10. The ear departed and having stayed away for a year returned and said, "How have you been able to live without me?" (They replied) "Like the deaf not hearing, but breathing with the breath, speaking with speech (the tongue), seeing with the eye, thinking with the mind. Thus (we lived)." The ear entered in.

11. The mind departed and having stayed away for a year returned and said, "How have you been able to live without me?" (They replied) "Like the children mindless but breathing with the breath, speaking with speech (the tongue), seeing with the eye, hearing with the ear. Thus (we lived)." The mind entered in.*

12. Now when breath was about to depart, tearing up the other senses, even as a spirited horse about to start might tear up the pegs to which he is tethered, they gathered round him and said, "Revered Sir, remain, you are the best of us, do not depart."

.    .    .

VI.8.   1. Then Uddalaka Aruni said to his son, Svetaketu, Learn from me, my dear, the true nature of sleep. When a person here sleeps, as it is called, then, my dear, he has reached pure being. He has gone to his own. Therefore they say he sleeps for he has gone to his own.

2. Just as a bird tied by a string, after flying in various directions without finding a resting place elsewhere settles down (at last) at the place where it is bound, so also the mind, my dear, after flying in various directions without finding a resting place elsewhere settles down in breath, for the mind, my dear, is bound to breath.**

.    .    .

VII.3.   2. He who meditates on mind as *Brahman* becomes independent so far as mind reaches, he who meditates on mind as *Brahman*. "Is there anything, Venerable Sir, greater than

* Compare page 56.
* * Compare page 56.

66

mind?" "Yes, there is something greater than mind." "Do, Venerable Sir, tell me that."

. . .

4.  1. Will, assuredly, is greater than mind. For when one wills, then one reflects, one utters speech, and then one utters it in name. The sacred hymns are included in name and sacred works in the sacred hymns.

    2. All these, verily, center in the will, have the will as their soul, abide in will. Heaven and earth were formed through will, air and ether were formed through will; water and heat were formed through will. Through their having been willed, rain becomes willed. Through rain having been willed, food becomes willed. Through food having been willed, living creatures are willed. Through living creatures having been willed, sacred hymns become willed. Through sacred hymns having been willed, sacred works become willed. Through sacred works having been willed, the world becomes willed. Through the world having been willed, everything becomes willed. Such is will. Meditate on will.

. . .

5.  1. Thought, assuredly, is more than will. Verily when one thinks, then he wills, then he reflects, then he utters speech and he utters it in name. The sacred hymns become one (are included) in name and sacred works in the sacred hymns.

    2. Verily, all these center in thought, have thought for their goal and abide in thought. Therefore, even if a man be possessed of much learning, but is unthinking, people say of him that he is nobody, whatever he may know. Verily, if he did know he would not be so unthinking. On the other hand, if he is thoughtful, even though he knows little, to him people are desirous of listening. Truly indeed thought is the center of all these, thought is their soul, thought is their support. Meditate on thought.

    3. He who meditates on thought as *Brahman*, he verily obtains the worlds he has thought, himself being permanent the permanent worlds, himself established, the established worlds, himself unwavering the unwavering worlds. As far as thought goes, so far is he independent, he who meditates on thought as *Brahman*. "Is there anything, Venerable Sir, greater than

thought?" "Yes, there is something greater than thought." "Do, Venerable Sir, tell me that."

.　　　.　　　.

6.　1. Contemplation, assuredly, is greater than thought. The earth contemplates as it were. The atmosphere contemplates as it were. The heaven contemplates as it were. The waters contemplate as it were, the mountains contemplate as it were. Gods and men contemplate as it were. Therefore he among men here attains greatness, he seems to have obtained a share of (the reward of) contemplation. Now the small people are quarrelsome, abusive, and slandering, the superior men seem to have obtained a share of (the reward of) contemplation, Meditate on contemplation.

2. He who meditates on contemplation as *Brahman*, so far as contemplation goes, so far is he independent, he who meditates on contemplation as *Brahman*. "Is there anything, Venerable Sir, greater than contemplation?" "Yes, there is something greater than contemplation." "Do, Venerable Sir, tell me that."

.　　　.　　　.

7.　1. Understanding, assuredly, is greater than contemplation. Verily, by understanding one understands the Rig-Veda . . . all this one understands just with understanding. Meditate on understanding.

2. He who meditates on understanding as *Brahman*, he verily, attains the worlds of understanding, of knowledge. As far as understanding goes, so far he is independent, he who meditates on understanding as *Brahman*. . . .

.　　　.　　　.

25.　2. Now next the instruction in regard to the self. The self indeed is below. The self is above. The self is behind. The self is in front. The self is to the south. The self is to the north. The self, indeed, is all this (world). Verily, he who sees this, who thinks this, who understands this, he has pleasure in the self, he has delight in the self, he has union in the self, he has joy in the self; he is independent (self-ruler); he has unlimited freedom in all worlds. But they who think differently from this are dependent on others (have others for their rulers). They have (live in) perishable worlds. In all worlds they cannot move at all (have no freedom).

. . .

26. 1. For him who sees this, who thinks this, and who understands this, life-breath springs from the self, hope from the self, memory from the self, ether from the self, heat from the self, water from the self, appearance and disappearance from the self, food from the self, strength from the self, understanding from the self, meditation from the self, thought from the self, determination from the self, mind from the self, speech from the self, name from the self, sacred hymns from the self, (sacred) works from the self, indeed all this (world) from the self.*

. . .

VIII.1. 5. . . . it (the self within) does not age with old age, it is not killed by the killing (of the body). That (and not the body) is the real city of Brahma. In it desires are contained. It is the self free from sin, free from old age, free from death, free from sorrow, free from hunger, free from thirst, whose desire is the real, whose thought is the real. . . .

6. As here on earth the world which is earned by work perishes, even so there the world which is earned by merit (derived from the performance of sacrifices) perishes. Those who depart hence without having found here the self and those real desires, for them there is no freedom in all the worlds. But

* If I could get within this changing I,
  This ever altering thing which yet persists,
  Keeping the features it is reckoned by,
  While each component atom breaks or twists,
  If, wandering past strange groups of shifting forms,
  Cells at their hidden marvels hard at work,
  Pale from much toil, or red from sudden storms,
  I might attain to where the Rulers lurk.
  If, pressing past the guards in those grey gates,
  The brain's most folded intertwisted shell,
  I might attain to that which alters fates,
  The King, the supreme self, the Master Cell,
  Then, in Man's earthly peak, I might behold
  The unearthly self beyond, unguessed, untold.
                                        —John Masefield

  (From *Good Friday and Other Poems.* Copyright 1916 by The Macmillan Company; renewed 1944 by John Masefield.)

those who depart hence, having found here the self and those real desires—for them in all worlds there is freedom.

7.   1. The self which is free from evil, free from old age, free from death, free from grief, free from hunger and thirst, whose desire is the real, whose thought is the real, he should be sought, him one should desire to understand. He who has found out and who understands that self, he obtains all worlds and all desires. Thus spoke Praja-pati.

## BRIHAD-ARANYAKA UPANISHAD

IV.3.   6. "When the sun has set, Yajnavalkya, and the moon has set, and the fire has gone out and speech has stopped, what light does a person here have?" "The self, indeed, is his light," said he, "for with the self, indeed, as the light, one sits, moves about, does one's work and returns."

7. "Which is the self?" "The person here who consists of knowledge among the senses, the light within the heart. He remaining the same, wanders along the two worlds seeming to think, seeming to move about. He on becoming asleep (getting into dream condition), transcends this world and the forms of death.

8. "Verily, this person, when he is born and obtains a body, becomes connected with evils. When he departs, on dying he leaves all evils behind.

9. "Verily, there are just two states of this person (the state of being in) this world and the state of being in the other world. There is an intermediate third state, that of being in sleep (dream). By standing in this intermediate state one sees both those states, of being in this world and of being in the other world. Now whatever the way is to the state of being in the other world, having obtained that way one sees both the evils (of this world) and the joys (of the other world). When he goes to sleep he takes along the material of this all-embracing world, himself tears it apart, himself builds it up; he sleeps (dreams) by his own brightness, by his own light. In that state the person becomes self-illuminated.

10. "There are no chariots there, nor animals to be yoked to them, no roads but he creates (projects from himself) chariots, animals to be yoked to them and roads. There are no joys

there, no pleasures, no delights, but he creates joys, pleasures, and delights. There are no ponds there, no lotus pools, no rivers, but he creates ponds, lotus pools, and rivers. He, indeed, is the agent (maker or creator).

11. "On this there are the following verses. Having struck down in sleep what belongs to the body, he himself sleepless looks down, on the sleeping (senses). Having taken to himself light he goes again to his place, the golden persons, the lonely swan (the one spirit).

12. "Guarding his low nest with the vital breath, the immortal moves out of the nest. That immortal one goes wherever he likes, the golden person, the lonely bird.

.　　.　　.

21. "This, verily, is his form which is free from craving, free from evils, free from fear. As a man when in the embrace of his beloved wife knows nothing without or within, so the person when in the embrace of the intelligent self knows nothing without or within. That, verily, is his form in which his desire is fulfilled, in which the self is his desire, in which he is without desire, free from any sorrow.

.　　.　　.

23. "Verily, when there (in the state of deep sleep) he does not see, he is, verily, seeing, though he does not see, for there is no cessation of the seeing of a seer, because of the imperishability (of the seer). There is not, however, a second, nothing else separate from him that he could see.

.　　.　　.

28. "Verily, when there (in the state of deep sleep) he does not think, he is, verily, thinking, though he does not think, for there is no cessation of the thinking of a thinker, because of the imperishability (of the thinker). There is not, however, a second, nothing else separate from him of which he could think.

.　　.　　.

32. "He becomes (transparent) like water, one, the seer without duality. This is the world of Brahma, Your Majesty." Thus did Yajnavalkya instruct (Janaka): "This is his highest goal;

this is his highest treasure; this is his highest world; this is his
greatest bliss. On a particle of this very bliss other creatures
live."

## PRACTICAL WISDOM

Much of the great tradition in Western psychology is the psychology of morals, customs, good judgment, the governing of a
good life. Indeed, such laws and principles pass over into practical
admonitions, such as those of the proverbs or adages which sum-
marize the "naïve psychology" of everyday assumptions about
human living. Such proverbs and monitions for the good life are
anchored on folkways or mores at one end, and on priestcraft and
statecraft, the wise governance of man, at the other end. In Aristotle's *Politics* there is much psychology which passes over into
wisdom in the governance of the state, and in Aristotle's *Rhetoric*
one finds the practical assumptions about human nature which the
orator must make use of if he is to convince a reluctant individual
or a pliable group. Psychology, in other words, pervades the various
assumptions and principles latent in every man's thought, and those
likewise which are crystallized into a wise man's rules of living; it
is not only in the technical psychology of the soul, or of the mind,
or of the learning process, that one finds "psychology." The same
general principles apply to the psychology of India. We have already seen that there are philosophical inquiries into the nature of
the mind; there are also in ancient India abundant principles
formulated for the use of those who govern. Just to give a bit of the
flavor, we draw here from the celebrated Laws of Manu (an early
post-Vedic verse treatise on legal topics), and from Kautilya's
Arthasastra (Kautilya was a legendary chancellor of about 300
B.C.; Arthasastra is a "Handbook of Wealth").

### The Laws of Manu*

II.    2. To act solely from a desire for rewards is not laudable, yet
an exemption from that desire is not (to be found) in this (world):

* Trans. by G. Bühler, from F. M. Müller (ed.), *Sacred Books of the East*,
Vol. XXV (Oxford: Clarendon Press, 1886).

for on (that) desire is grounded the study of the Veda and the performance of the actions, prescribed by the Veda.

4. Not a single act here (below) appears ever to be done by a man free from desire; for whatever (man) does, it is (the result of) the impulse of desire.*

98. That man may be considered to have (really) subdued his organs, who on hearing and touching and seeing, on tasting and smelling (anything) neither rejoices nor repines.

145. The teacher (*akarya*) is ten times more venerable than a sub-teacher (*upadhyaya*), the father a hundred times more than the teacher, but the mother a thousand times more than the father.

VI.     48. Against an angry man let him not in return show anger, let him bless when he is cursed, and let him not utter speech, devoid of truth, scattered at the seven gates.

92. Contentment, forgiveness, self-control, abstention from un-righteously appropriating anything, (obedience to the rules of) purification, coercion of the organs, wisdom, knowledge (of the supreme Soul), truthfulness, and abstention from anger, (form) the tenfold law.

## Kautilya's Arthasastra**

PUNISHMENT

Whoever imposes severe punishment becomes repulsive to the people; while he who awards mild punishment becomes contemptible. But whoever imposes punishment as deserved becomes respectable. For punishment (*danda*), when awarded with due consideration, makes the people devoted to righteousness and to work productive of wealth and enjoyment; while punishment, when ill-awarded under the influence of greed and anger or owing to ignorance, excites fury even among hermits and ascetics dwelling in forests, not to speak of householders.

But when the law of punishment is kept in abeyance, it gives rise to such disorder as is implied in the proverb of fishes (*matsyanyayamud-havayati*), for in the absence of a magistrate (*dandadharabhave*), the strong will swallow the weak; but under his protection the weak resist the strong.

*Danda*, punishment, which alone can procure safety and security of life is, in its turn, dependent on discipline (*vinaya*).

Discipline is of two kinds: artificial and natural; for instruction (*kriya*) can render only a docile being conformable to the rules of

* Compare page 71.
** From R. Shamasasty (trans.), *Kautilya's Arthasastra* (2nd ed.; Mysore: Wesleyan Missionary Press, 1923).

discipline, and not an undocile being (*adravyam*). The study of sciences can tame only those who are possessed of such mental faculties as obedience, hearing, grasping, retentive memory, discrimination, inference, and deliberation, but not others devoid of such faculties.

## CREATION OF MINISTERS

"The king," says Bharadvaja, "shall employ his classmates as his ministers; for they can be trusted by him inasmuch as he has personal knowledge of their honesty and capacity."

"No," says Visalaksha, "for, as they have been his playmates as well, they would despise him. But he shall employ as ministers those whose secrets, possessed of in common, are well known to him. Possessed of habits and defects in common with the king, they would never hurt him lest he would betray their secrets."

"Common is this fear," says Parasara, "for under the fear of betrayal of his own secrets, the king may also follow them in their good and bad acts.

"Under the control of as many persons as are made aware by the king of his own secrets, might he place himself in all humility by that disclosure. Hence he shall employ as ministers those who have proved faithful to him under difficulties fatal to life and are of tried devotion."

"No," says Pisuna, "for this is devotion, but not intelligence (*buddhigunah*). He shall appoint as ministers those who, when employed on financial matters, show as much as, or more than, the fixed revenue, and are thus of tried ability."

"No," says Kaunapadanta, "for such persons are devoid of other ministerial qualifications; he shall, therefore, employ as ministers those whose fathers and grandfathers had been ministers before; such persons, in virtue of their knowledge of past events and of an established relationship with the king, will, though offended, never desert him; for such faithfulness is seen even among dumb animals; cows, for example, stand aside from strange cows and ever keep company with accustomed herds."

"No," says Vatavyadhi, "for such persons, having acquired complete dominion over the king, begin to play themselves as the king. Hence he shall employ as ministers such new persons as are proficient in the science of polity. It is such new persons who will regard the king as the real scepter-bearer (*dandadhara*) and dare not offend him."

"No," says the son of Bahudanti (a woman); "for a man possessed of only theoretical knowledge, and having no experience of practical politics, is likely to commit serious blunders when engaged in actual works. Hence he shall employ as ministers such as are born of high family and possessed of wisdom, purity of purpose, bravery and loyal feelings, inasmuch as ministerial appointments shall purely depend on qualifications."

"This," says Kautilya, "is satisfactory in all respects; for a man's ability is inferred from his capacity shown in work. And in accordance with the difference in the working capacity.

"Of these qualifications, native birth and influential position shall be ascertained from reliable persons; educational qualifications (*silpa*) from professors of equal learning; theoretical and practical knowledge, foresight, retentive memory, and affability shall be tested from successful application in works; eloquence, skillfulness, and flashing intelligence from power shown in narrating stories (*kathayogeshu*, i.e. in conversation); endurance, enthusiasm, and bravery in troubles; purity of life, friendly disposition, and loyal devotion by frequent association; conduct, strength, health, dignity, and freedom from indolence and fickleminded-ness shall be ascertained from their intimate friends; and affectionate and philanthrophic nature by personal experience."

COUNCIL MEETING

"No deliberation," says Visalaksha, "made by a single person will be successful; the nature of the work which a sovereign has to do is to be inferred from the consideration of both the visible and invisible causes. The perception of what is not, or cannot be seen, the conclusive decision of whatever is seen, the clearance of doubts as to whatever is susceptible of two opinions, and the inference of the whole when only a part is seen—all this is possible of decision only by ministers. Hence he shall sit at deliberation with persons of wide intellect."

MORALE

Which is to be marched against—an enemy whose subjects are impoverished and greedy or an enemy whose subjects are being oppressed?

My teacher says that the conqueror should march against that enemy whose subjects are impoverished and greedy, for impoverished and greedy subjects suffer themselves to be won over to the other side by intrigue, and are easily excited. But not so the oppressed subjects whose wrath can be pacified by punishing the chief men (of the state).

Not so, says Kautilya: for though impoverished and greedy they are loyal to their master and are ready to stand for his cause and to defeat any intrigue against him; for it is in loyalty that all other good qualities have their strength. Hence the conqueror should march against the enemy whose subjects are oppressed.

### THE BHAGAVAD-GITA

The introductory material on Satya (pages 39 ff., above), the texts of and commentaries on the Upanishads, and our brief selections from the Laws of Manu and Kautilya's Arthasastra give the Hindu message in prose, and for the most part in rather abstract prose. It remains to offer the message in poetic form: indeed, in one of the great poems of all time, in which the recurrent idea of Hindu speculative thought takes a personal form that is expressive of a great soul in doubt, striving, in a dialogue with Deity, to make sense of the eternal human predicament. This poem or "song," the Bhagavad-Gita, is embedded in the great epic poem the Mahabharata, a song which has been the prized jewel of countless devout Hindus, and admired and beloved everywhere that religious poetry is revered.

As the warrior prepares to lead his host against the vast enemy armies before him, his knees shake at the contemplation of bloodshed in a war of brothers, and the futility and wretchedness of human life in which such warfare occurs. His charioteer, who is the divine Krishna in human form, explains that spiritual reality continues regardless of weapons, wounds, and destruction; the soul goes on following a spiritual career which is neither arrested nor thrown from its course by the sheer physical events of life or death. It was this that prompted from Ralph Waldo Emerson, in his poem "Brahma," the beautiful and frequently quoted lines:

> If the red slayer think he slays,
>   or if the slain think he is slain,
> They know not well the subtle ways
>   I keep, and pass, and turn again . . .
> They reckon ill who leave me out,
>   When me they fly, I am the wings;
> I am the doubter and the doubt,
>   And I the hymn the Brahmin sings.

The Bhagavad-Gita is a richly poetic dialogue between a great king leading his hosts to battle and his charioteer, a deity in human form. As the moment of the great attack and terrible carnage approaches, and the king shudders at the death of brothers at the hands of brothers, somehow the immediate world of the senses is drawn away and eternal realities about the nature of man's inner self take visible shape. In a sense every Indian knows the Bhagavad-Gita, for the Mahabharata is deeply in the Indian tradition, even more deeply than Homer is in the Western tradition. Actually, however, India has in recent decades been progressively rediscovering her past. In Gandhi's *Autobiography* we read how, as a young law student in England, he represented in himself a struggle between an India beginning to reassert itself and an India craving contact with the West. In the long search for his own identity and his own understanding of the Indian tradition, he came upon the Bhagavad-Gita and was overwhelmed by it. It seems hardly possible to the Western reader that Gandhi should have been unfamiliar with the Bhagavad-Gita, so central is its place in the Indian tradition; it seems likewise impossible that one searching contact with Western economics and politics, as they appear in Western law, should have been carried away by the exquisite beauty and power of a religious, indeed a profoundly mystical, message. That is, however, clearly what happened. The Bhagavad-Gita represents the poetic version of the Upanishadic search for meaning, and the search for meaning lies not in external relations, but in a changeless inner self. This must, however, be pursued with some regard for difficult technical issues, and we have asked Professor Kuppuswamy to give us the following commentary on the Bhagavad-Gita at this point.

### The Indriyas, the Manas, the Buddhi, and the Self

The author of the Bhagavad-Gita speaks of the self, intellect (*buddhi*), the ten senses (*indriyas*)—five sensory and five pertaining to action—and the mind (*manas*), as the basic constituents of the personality. The senses, the mind, and the intelligence obstruct the wisdom of the self. Consequently, one has to control the senses, which are destroyers of wisdom and discrimination. We also find here the hierarchic view of the various aspects of personality, according to which the mind is greater than the senses, intellect (*buddhi*) is greater than the mind, and the self is greater than intellect. A man can have mastery over his self only when he has conquered his senses and when he acts without being *involved* in action. There is also a distinction between two levels of personality: the lower and the higher. The lower consists of the ten senses, the mind, and the intellect (*prakriti*), and the higher consists of the self (*purusha*).

Wisdom consists in the discrimination between the field and the knower of the field, or knowledge and the object of knowledge. The body is the field to be known and the self is the knower. Here again the body is explained as consisting of the five gross elements, the five objects of sense, the ten *indriyas*, the *manas*, the *buddhi*, and the self-sense. Desire and hatred, pleasure and pain, intelligence and steadfastness, are the modifications of nature (*prakriti*).

All the sensations and feelings of pleasure and pain arise from the contact of the sense organs with objects. Thus, one must realize that these sensations and feelings are transitory and one must learn to endure them when one experiences them. A man of wisdom, then, is one who is in control of these sensations and feelings, and who remains the same when he is experiencing them. One's aim should be to withdraw one's senses from the objects of the senses so that one's intelligence is firmly set. It is, however, realized that the mastery of the senses is a very difficult task. A person who has attained a mastery over the senses will be able to move among objects of sense without either attachment or aversion.

### The Self

The self appears to have all the qualities of the senses and yet is something that cannot be perceived. It is looked upon as enjoying, as well as supporting, all organs of sense and action, but it is unattached. It is free from the three *gunas*—*sattva*, *rajas*, and *tamas*. The self is also regarded as possessing contradictory qualities. It—or he—is within every living being, but he is also outside. He is moving, but he is also stationary. He is far away, yet near at hand. He is indivisible, but he is also to be found divided among beings. Two parts, *prakriti* and *purusha*, are recognized. All changes in the individual are attributed to changes in *prakriti*. The self is looked upon as the cause of the experiences of pleasure and pain. The ultimate aim, according to the Bhagavad-Gita, is the knowledge of the self and of the *prakriti*, or nature, together with all its modes. It is also stated that there are three methods of knowing the self: the path of meditation, the path of knowledge, and the path of works. It consists in the realization that all our actions are done by the *prakriti*. The individual himself is not the doer; he is only a witness. The self is beyond the reach of every weapon, or fire, or water, or wind. The self is unmanifest, unthinkable, and unchanging, and the individual should endeavor to know this. It is also stated that the self enjoys objects through the sense organs and the mind. Deluded individuals identify the self with the mind and the sense organs, but wise men are able to see the difference. There are two persons in this world, the one which is perishable (the aspect connected with the *prakriti*) and the one which is imperishable (the self). The man with delusions recognizes only the first aspect, whereas the man with wisdom recognizes the second.

## The Three Gunas

One of the most important contributions of the Bhagavad-Gita is the doctrine of the three *gunas*. It asserts that every living being behaves on account of impulses born of nature. According to Indian thought, there are three *gunas*, or aspects of physical reality—*sattva, rajas,* and *tamas*. In a general way *sattva* signifies something pure and fine, *rajas* stands for activity, and *tamas* for what is stolid and offers resistance. Thus this view tries to isolate the three opposing elements in nature. It is also held that these three aspects act together; most often the illustration of the flame is used to explain their combined action: flame is the result of the cooperation between the wick, the oil, and the fire. These three constituents are to be found in every object in the universe. This idea clearly represents a recognition of the physiological basis of behavior. The same idea is expressed in another way: No human being can live without action; the very maintenance of the physical life is not possible without action.

All of Chapter XVII is devoted to the analysis of human action on the basis of the doctrine of the three *gunas*. The faith of every individual depends upon his nature. The concept has also been used to classify and analyze various kinds of human activities. It is stated, for example, that people in whom *sattva* predominates will worship God, while those in whom *rajas* predominates will worship the demigods and the demons, and those in whom *tamas* predominates will worship ghosts and spirits. The person in whom *sattva* predominates offers sacrifice according to scriptural laws, while those with *rajas* offer either with the expectation of reward or for the sake of display, and the third group will do so without any kind of faith. A person in whom *sattva* predominates will be pure, upright, content, and nonviolent in his action; truthful, pleasant, and beneficial in speech; serene, gentle, silent, and capable of self-control. Those in whom *rajas* predominates will be passionate and their action will be motivated to gain respect, honor, and reverence. Those in whom *tamas* predominates will be foolish and obstinate and will either torture themselves or cause injury to others.

A person who is bewildered and who is filled with egoism imagines that he is the doer, whereas in reality anything that he does is due merely to the three modes of nature. When a person understands this, it is stated, he will not be attached to his actions. On the other hand, the others, because of their ignorance of the operation of the modes of nature, will become attached to their actions. Another verse asserts that the man of knowledge will act in accordance with his own nature and it questions both the possibility and the beneficence of repression.

Thus, according to the Bhagavad-Gita, all behavior or action, or *karma* (the necessary consequences of action), arises from the three *gunas*, the modes of nature. An individual who identifies himself with his physiological-social being becomes attached to the work that he does, and suffers. Consequently, an individual whose actions are free from desire and egoism, who is concentrating on the self rather than on action, will be free.

There is a famous verse in the Bhagavad-Gita which has occasioned a good deal of controversy. It asserts that the man who does something that is prescribed by his own nature is better than the man who tries to do something on the basis of imitation. The traditional interpretation of this verse is linked with the caste system. From the modern point of view, it would seem that the idea has a deeper significance. It may be stated that an individual who engages himself in the activities in which he has gained competence, and does not try to imitate the activities of others merely because they are more attractive or of a higher economic benefit or social prestige, is one who can be efficient and effective.

The Bhagavad-Gita clearly recognizes that behavior is physico-psychological: "Everyone is made to act helplessly by the impulses born of nature." It also states that no person can remain even for a moment without performing some action. If action is inevitable, are there no levels of actions? The individual who controls the cognitive as well as the executive organs and the mind, and who acts without attachment, is superior.

Our attention is also drawn to the social norms; one's activities are modeled on what a great man does. An enlightened person should continue to engage himself in the normal social activities, for he should act as a model for other people; he is also warned that he should not unsettle the minds of ignorant persons who are attached to their actions. Another section contains a discussion of the actions that run contrary to the social norms. It asks: How is a man compelled against his will to commit sins? It is asserted that the real enemies are craving and wrath; so it is that a person who is overpowered by his passions commits sin. In every person wisdom is enveloped by passion; one yields to his desires and forsakes reality—that is, ideals as well as the social norms.

There is a long discussion about the problem of action and inaction. One must understand the difference between action, wrong action, and inaction. A person who sees "action in inaction" and "inaction in action" is the most competent person. We may distinguish three aspects. First of all, a highly competent man can perform a most complicated task with the least sense of effort, so that his action looks as if it were inaction. Secondly, there is an aspect of attachment to the action. When an individual is attached to his action he is neither competent nor satisfied. Finally, there is the aspect of the fruits of action. A person who is constantly thinking of the consequences of his actions is not able to do his work in an efficient manner. It is also asserted that a man is wise if he undertakes action without desires and with full knowledge of what he is doing. As the Bhagavad-Gita puts it, he is one "whose works are burned up in the fire of wisdom." Such an individual may continue to do his work, but he is unaffected either by a sense of effort or by a sense of frustration.

Chapter V sets forth the same idea about action. There is the famous question about the difference between renunciation of action and unselfish performance of action. A person who is well trained and who has conquered his cognitive as well as his executive organs, and who has attained a mastery over his self, is one who is not affected by his work though he is constantly doing his work.

One of the most famous verses of the Bhagavad-Gita states that an individual has a right only to act and not to the fruits of his actions. However, though one gives up the fruits of action, he should not develop an attachment to inaction. The same idea is expressed in a slightly different way in the statement that an individual is continually engaged in action. All one's actions should be dedicated. If one performs his action out of a self-conceit, then he will perish. According to the Bhagavad-Gita, the wise man is not the one who has abstained from action but the one who has given up the fruits of action.

There are five factors that enter into the accomplishment of an action: (1) the seat of mind, (2) the agent, (3) the instrument used, (4) the efforts made, and (5) providence.

Whether it is a bodily act or speech, or mere desiring or willing, all five factors are involved, whether the action is a right or a wrong one. So a person who imagines that he is the sole agent of his actions is perverse of mind, according to the Bhagavad-Gita. On the other hand, the one who realizes that his actions are the product of these five factors is an enlightened person. Similarly, all actions involve knowledge: the object of knowledge and the knowing subject, on the one hand, and the instrument of the action and the agent, on the other.

All actions are classified, as we have seen, according to the three *gunas*. Those which are regarded as obligatory and which are performed without attachment, without love or hatred, without any desire for the fruit, such actions express *sattva*. Those actions which involve a good deal of effort and through which a person seeks to gratify his desires or to enhance his prestige are classified as the expression of *rajas*. Finally, actions which are done through ignorance, without regard to consequences, which may lead to loss or injury, or actions which do not take into account one's ability and limitations, such actions are classified as the expression of *tamas*.

### Preparation to Attain Liberation

In almost every chapter there are references to how an individual should equip himself in order to attain liberation. This preparation is long and arduous. However, though it is difficult, no effort in this direction is ever lost. Small gains will ultimately produce satisfactory results. In this connection it is stated that singleness of purpose is most important. Normally we are irresolute; so are our desires; our thoughts are many and endless. One of the important tasks, therefore, is to bring about singleness of purpose, so that a wavering mind becomes resolute.

Among the various means of attaining liberation, according to the Bhagavad-Gita, are the following: One should realize that one's thoughts, speech, and action arise out of the needs of the body and the mind. In other words, one must understand that he is above this threefold nature of *sattva, rajas*, and *tamas*. A person should also be free from the pairs of opposites. He should be cordial to his foe as well as to his friend, indifferent to good as well as evil reputation; he should be the same in pleasure or pain. It is often stated that one should bear cold as well as heat; similarly, one should neither rejoice nor grieve. It is only by renouncing both good and evil and by renouncing all attachment that one can hope to develop. Yet another means for the attainment of liberation is the control over the organs, cognitive as well as executive, *indriyanigraha*. When a person subdues his senses, he gains wisdom; when he gains wisdom, he attains peace of mind. Yet another characteristic is freedom from the influence of doubt and anxiety. Faith is an essential ingredient for the attainment of liberation. A person who doubts will always be miserable. Doubt is born of ignorance and it can be cut apart by the sword of wisdom.

One of the most remarkable verses of the Bhagavad-Gita says that a man can lift himself to a higher level by his own efforts. It is stated that it is the self alone which is a friend of the self. Similarly, it is the self alone which is the enemy of the self. Thus it is only through the conquest of one's self that it is possible to rise up. Chapter VI, which is traditionally considered one of the most important, contains many statements designed to help an individual to prepare himself for a better life. An individual, according to one verse, should restrain all the senses; he should be even-minded in all conditions; and he should rejoice in the welfare of all creatures.

A man who gains wisdom and can distinguish between the field and the knower of the field is able to attain liberation. A person who is established in *sattva* (goodness and equanimity), rises upward, while a person in whom *rajas* (passion) predominates will be in the middle region, and the person in whom *tamas* (dullness) predominates sinks downward. Thus the important aspect here is the ability to discriminate between the self and the body-mind complex that is responsible for all our actions through the three modes of *sattva, rajas,* and *tamas.* Another important aspect is the necessity for freedom from pride and delusion and the importance of overcoming desires and transcending the dualities of pleasure and pain. This is attained through nonattachment.

### Characteristics of Wise Men

The Bhagavad-Gita is replete with references to the attributes of the behavior of wise, or liberated, men. One of the important characteristics of a man of maturity is that he does not long for sensuous contacts with objects. He brings his senses, cognitive as well as executive, under control, and transcends duality. In this way he can move among the objects of sense without any danger. Such a man, moreover, will not experience any sorrow; he will be able to put away all desires; he will be free from egoism and a sense of "mineness."

On the positive side, he will enjoy contentment and will experience delight in himself. Consequently, he will perform all actions without any attachment. In two remarkable verses the Bhagavad-Gita also speaks of freedom from passion and longing, fear and anger. The wise man will not only have a mind that is subdued and fully under control; he will also be able to look upon a clod of earth and a piece of gold with equal indifference. The Bhagavad-Gita also describes how the wise man takes his food, rest, and exercise.

The Bhagavad-Gita recognizes individual differences among human beings in their striving for perfection. It points out the well-known paths of meditation, knowledge, and works. In fact, one of the charges directed against the author of the Bhagavad-Gita is that he has indicated a number of paths toward self-realization. Sometimes these paths are contradictory; but it must be realized that he was trying to solve the problem of individual differences and variations in aptitude and temperament.

There is also a description of the attributes of the four classes of human beings—the scholar, the warrior, the businessman, and the worker—that is, the seeker after knowledge, the seeker after power, the seeker after wealth, and the rest of mankind, who do not have any definite goals.

Imagine, then, the scene: The vast hosts of brothers and kinsmen deployed upon the battlefield, and about to be hurled forth against one another. The storyteller, however, at the critical moment yields to the philosopher, just as we, in the reading of the history of India, turn away from the stories of struggle and dominion, and find ourselves lost in the world of contemplation. The leader of the hosts, giddy with sorrow and barely able to face his terrible duty, is drawn on by his divine charioteer into the whole cycle of problems about the self and the world which we have already encountered in the earlier Indian classics.

Homer, too, wanders in the midst of the struggle of Achilles and Agamemnon in the sublime and terrible story of the avenging of Patroclus by the death of Hector, and allows himself little images of the swarming of bees and the swaying of grain in the wind. But in the Western tradition these digressions are brief, and must be justified as adornments to the tale. In the Indian poetic imagination the reality is the digression itself, and the hosts must wait through the hours in which eternal matters are poetically contemplated.

### From the Bhagavad-Gita*

I.  26. There the son of Prtha saw stationed
         Fathers and grandsires,
         Teachers, uncles, brothers,
         Sons, grandsons, and comrades too,

* All quotations from the Bhagavad-Gita are from Franklin Edgerton (trans.), *The Bhagavad Gita* (Cambridge, Mass.: Harvard University Press, 1952). Copyright 1944 by the President and Fellows of Harvard College.

84

27. Fathers-in-law and friends as well,
      In both the two armies.
    The son of Kunti, seeing them,
      All his kinsmen arrayed,

28. Filled with utmost compassion,
      Despondent, spoke these words:
    Seeing my own kinsfolk here, Krsna,
      That have drawn near eager to fight,

29. My limbs sink down,
      And my mouth becomes parched,
    And there is trembling in my body,
      And my hair stands on end.

30. (The bow) Gandiva falls from my hand,
      And my skin, too, is burning,
    And I cannot stand still,
      And my mind seems to wander.

                    .        .        .

38. Even if they do not see,
      Because their intelligence is destroyed by greed,
    The sin caused by destruction of family,
      And the crime involved in injury to a friend,

39. How should we not know enough
      To turn back from this wickedness,
    The sin caused by destruction of family
      Perceiving, O Janardana?

                    .        .        .

47. Thus speaking Arjuna in the battle
      Sat down in the box of the car,
    Letting fall his bow and arrows,
      His heart smitten with grief.

                    .        .        .

II.    7. My very being afflicted with the taint of weak compassion,
          I ask Thee, my mind bewildered as to the right:
        Which were better, that tell me definitely;
            I am Thy pupil, teach me that have come to Thee (for
              instruction).

8. For I see not what would dispel my
   Grief, the witherer of the senses,
   If I attained on earth rivalless, prosperous
   Kingship, and even overlordship of the gods.

. . .

The Blessed One said:

11. Thou hast mourned those who should not be mourned,
    And (yet) thou speakest words about wisdom!
    Dead and living men
    The (truly) learned do not mourn.

. . .

14. But contacts with matter, son of Kunti,
    Cause cold and heat, pleasure and pain;
    They come and go, and are impermanent;
    Put up with them, son of Bharata!

15. For whom these (contacts) do not cause to waver.
    The man, O bull of men,
    To whom pain and pleasure are alike, the wise,
    He is fit for immortality.

. . .

23. Swords cut him not,
    Fire burns him not,
    Water wets him not,
    Wind dries him not.

24. Not to be cut is he, not to be burned is he,
    Not to be wet nor yet dried;
    Eternal, omnipresent, fixed,
    Immovable, everlasting is he.

25. Unmanifest he, unthinkable he,
    Unchangeable he is declared to be;
    Therefore knowing him thus
    Thou shouldst not mourn him.

. . .

45. The Vedas have the three Strands (of matter) as their scope;
    Be thou free from the three Strands, Arjuna,
    Free from the pairs (of opposites), eternally fixed in goodness,
    Free from acquisition and possession, self-possessed.

. . .

86

47. On action alone be thy interest,
     Never on its fruits;
     Let not the fruits of action be thy motive,
     Nor be thy attachment to inaction.

48. Abiding in discipline perform actions,
     Abandoning attachment, Dhanamjaya,
     Being indifferent to success or failure;
     Discipline is defined as indifference.

49. For action is far inferior
     To discipline of mental attitude, Dhanamjaya.
     In the mental attitude seek thy (religious) refuge;
     Wretched are those whose motive is the fruit (of action).

50. The disciplined in mental attitude leaves behind in this world
     Both good and evil deeds.
     Therefore discipline thyself unto discipline;
     Discipline in actions is weal.

51. For the disciplined in mental attitude, action-produced
     Fruit abandoning, the intelligent ones,
     Freed from the bondage of rebirth,
     Go to the place that is free from illness.


                      .        .        .


        Arjuna said:
54. What is the description of the man of stabilized mentality,
     That is fixed in concentration, Kesava?
     How might the man of stabilized mentality speak,
     How might he sit, how walk?

        The Blessed One said:
55. When he abandons desires,
     All that are in the mind, son of Prtha,
     Finding contentment by himself in the self alone,
     Then he is called of stabilized mentality.

56. When his mind is not perturbed in sorrows,
     And he has lost desire for joys,
     His longing, fear, and wrath departed,
     He is called a stable-minded holy man.

57. Who has no desire toward any thing,
     And getting this or that good or evil
     Neither delights in it nor loathes it,
     His mentality is stabilized.

58. And when he withdraws,
    As a tortoise his limbs from all sides,
    His senses from the objects of sense,
    His mentality is stabilized.

        .        .        .

63. From wrath comes infatuation,
    From infatuation loss of memory;
    From loss of memory, loss of mind;
    From loss of mind he perishes.

64. But with desire-and-loathing-severed
    Senses acting on the objects of sense,
    With (senses) self-controlled, he, governing his self,
    Goes unto tranquillity.

        .        .        .

67. For the senses are roving,
    And when the thought-organ is directed after them,
    It carries away his mentality,
    As wind a ship on the water.

68. Therefore whosoever, great-armed one,
    Has withdrawn on all sides
    The senses from the objects of sense,
    His mentality is stabilized.

69. What is night for all beings,
    Therein the man of restraint is awake;
    Wherein (other) beings are awake,
    That is night for the sage of vision.

70. It is ever being filled, and (yet) its foundation remains un-
        moved
    The sea: just as waters enter it,
    Whom all desires enter in that same way
    He attains peace; not the man who lusts after desires.

71. Abandoning all desires, what
    Man moves free from longing,
    Without self-interest and egotism,
    He goes to peace.

72. This is the fixation that is Brahmanic, son of Prtha;
    Having attained it he is not (again) confused.
    Abiding in it even at the time of death,
    He goes to Brahman-nirvana.

.    .    .

The Blessed One said:

III.    3. In this world a twofold basis (of religion)
Has been declared by Me of old, blameless one:
By the discipline of knowledge of the followers of reason-
method,
And by the discipline of action of the followers of discipline-
method

.    .    .

5. For no one even for a moment
Remains at all without performing actions;
For he is made to perform action willy-nilly,
Every one is, by the Strands that spring from material
nature.

6. Restraining the action-senses
Who sits pondering with his thought-organ
On the objects of sense, with deluded soul,
He is called a hypocrite.

7. But whoso the senses with the thought-organ
Controlling, O Arjuna, undertakes
Discipline of action with the action-senses,
Unattached (to the fruits of action), he is superior.

8. Perform thou action that is (religiously) required;
For action is better than inaction.
And even the maintenance of the body for thee
Can not succeed without action.

.    .    .

16. The wheel thus set in motion
Who does not keep turning in this world,
Malignant, delighting in the senses,
He lives in vain, son of Prtha.

17. But who takes delight in the self alone,
The man who finds contentment in the self,
And satisfaction only in the self,
For him there is found (in effect) no action to perform.

18. He has no interest whatever in action done,
Nor any in action not done in this world,
Nor has he in reference to all beings
Any dependence of interest.

19. Therefore unattached ever
     Perform action that must be done;
   For performing action without attachment
     Man attains the highest.

      .      .      .

21. Whatsoever the noblest does,
     Just that in every case other folk (do);
   What he makes his standard,
     That the world follows.

      .      .      .

25. Fools, attached to action,
     As they act, son of Bharata,
   So the wise man should act (but) unattached,
     Seeking to effect the control of the world.

26. Let him not cause confusion of mind
     In ignorant folk who are attached to action;
   He should let them enjoy all actions,
     The wise man, (himself) acting disciplined.

27. Performed by material nature's
     Strands are actions, altogether;
   He whose soul is deluded by the I-faculty
     Imagines "I am the agent."

28. But he who knows the truth, great-armed one,
     About the separation (of the soul) from both the Strands
          and action,
   "The Strands act upon the Strands"—
     Knowing this, is not attached (to actions).

29. Deluded by the Strands of material nature,
     Men are attached to the actions of the Strands.
   These dull folk of imperfect knowledge
     The man of perfect knowledge should not disturb.

30. On Me all actions
     Casting, with mind of the over-soul,
   Being free from longing and from selfishness,
     Fight, casting off thy fever.

      .      .      .

33. One acts in conformity with his own
     Material nature—even the wise man;
     Beings follow (their own) nature;
     What will restraint accomplish?

34. Of (every) sense, upon the objects of (that) sense
     Longing and loathing are fixed;
     One must not come under control of those two,
     For they are his two enemies.

35. Better one's own duty, (tho) imperfect,
     Than another's duty well performed;
     Better death in (doing) one's own duty;
     Another's duty brings danger.


         .      .      .


IV.  18. Who sees inaction in action,
     And action in inaction,
     He is enlightened among men;
     He does all actions, disciplined.

19. All whose undertakings
     Are free from desire and purpose,
     His actions burned up in the fire of knowledge,
     Him the wise call the man of learning.

20. Abandoning attachment to the fruits of action,
     Constantly content, independent,
     Even when he sets out upon action,
     He yet does (in effect) nothing whatsoever.

21. Free from wishes, with mind and soul restrained,
     Abandoning all possessions,
     Action with the body alone
     Performing, he attains no guilt.

22. Content with getting what comes by chance,
     Passed beyond the pairs (of opposites), free from jealousy,
     Indifferent to success and failure,
     Even acting, he is not bound.

23. Rid of attachment, freed,
     His mind fixed in knowledge,
     Doing acts for worship (only), his action
     All melts away.

.  .  .

38. For not like unto knowledge
Is any purifier found in this world.
This the man perfected in discipline himself
In time finds in himself.

.  .  .

The Blessed One said:

V.  2. Renunciation and discipline of action
Both lead to supreme weal.
But of these two, rather than renunciation of action,
Discipline of action is superior.

3. He is to be recognized as (in effect) forever renouncing
(action),
Who neither loathes nor craves;
For he that is free from the pairs (of opposites), great-armed
one,
Is easily freed from bondage (otherwise caused by actions).

.  .  .

7. Disciplined in discipline, with purified self,
Self-subdued, with senses overcome,
His self become (one with) the self of all beings,
Even acting, he is not stained.

8. "I am (in effect) doing nothing at all!"—so
The disciplined man should think, knowing the truth,
When he sees, hears, touches, smells,
Eats, walks, sleeps, breathes,

9. Talks, evacuates, grasps,
Opens and shuts his eyes;
"The senses (only) on the objects of sense
Are operating"—holding fast to this thought.

10. Casting (all) actions upon Brahman,
Whoso acts abandoning attachment,
Evil does not cleave to him,
As water (does not cleave) to a lotus-leaf.

11. With the body, the thought-organ, the intelligence,
And also with the senses alone,
Disciplined men perform action,
Abandoning attachment, unto self-purification.

12. The disciplined man, abandoning the fruit of actions,
    Attains abiding peace;
    The undisciplined, by action due to desire,
    Attached to the fruit (of action), is bound.

13. All actions with the thought-organ
    Renouncing, he sits happily, in control,
    The embodied (soul), in the citadel of nine gates,
    Not in the least acting nor causing to act.

     .     .     .

20. He will not rejoice on attaining the pleasant,
    Nor repine on attaining the unpleasant;
    With stabilized mentality, unbewildered,
    Knowing Brahman, he is fixed in Brahman.

21. With self unattached to outside contacts,
    When he finds happiness in the self,
    He, his self disciplined in Brahman-discipline,
    Attains imperishable bliss.

22. For the enjoyments that spring from (outside) contacts
    Are nothing but sources of misery;
    They have beginning and end, son of Kunti;
    The wise man takes no delight in them.

23. Who can control right in this life,
    Before being freed from the body,
    The excitement that springs from desire and wrath,
    He is disciplined, he the happy man.

24. Who finds his happiness within, his joy within,
    And likewise his light only within,
    That disciplined man to Brahman-nirvana
    Goes, having become Brahman.

25. Brahman-nirvana is won
    By the seers whose sins are destroyed,
    Whose doubts are cleft, whose souls are controlled,
    Who delight in the welfare of all beings.

26. To those who have put off desire and wrath,
    Religious men whose minds are controlled,
    Close at hand Brahman-nirvana
    Comes, to knowers of the self.

     .     .     .

VI. 3. For the sage that desires to mount to discipline
Action is called the means;
For the same man when he has mounted to discipline
Quiescence is called the means.

4. For when not to the objects of sense
Nor to actions is he attached,
Renouncing all purpose,
Then he is said to have mounted to discipline.

5. One should lift up the self by the self,
And should not let the self down;
For the self is the self's only friend,
And the self is the self's only enemy.

6. The self is a friend to the self
By which self the very self is subdued;
But to him that does not possess the self, in enmity
Will abide his very self, like an enemy.

7. Of the self-subdued, pacified man,
The supreme self remains concentrated (in absorption),
In cold and heat, pleasure and pain,
Likewise in honor and disgrace.

8. His self satiated with theoretical and practical knowledge,
Immovable, with subdued senses,
The possessor of discipline is called (truly) disciplined,
To whom clods, stones, and gold are all one.

.  .  .

11. In a clean place establishing
A steady seat for himself,
That is neither too high nor too low,
Covered with a cloth, a skin, and kusa-grass,

12. There fixing the thought-organ on a single object,
Restraining the activity of his mind and senses,
Sitting on the seat, let him practice
Discipline unto self-purification.

13. Even body, head, and neck
Holding motionless, (keeping himself) steady,
Gazing at the tip of his own nose,
And not looking in any direction,

14. With tranquil soul, rid of fear,
     Abiding in the vow of chastity,
     Controlling the mind, his thoughts on Me,
     Let him sit disciplined, absorbed in Me.

15. Thus ever disciplining himself,
     The man of discipline, with controlled mind,
     To peace that culminates in nirvana,
     And rests in Me, attains.

.    .    .

19. As a lamp stationed in a windless place
     Flickers not, this image is recorded
     Of the disciplined man controlled in thought,
     Practicing discipline of the self.

It is a sort of sacrilege to delete parts of the Bhagavad-Gita, as we have done. The song has become like a sacred body, or like a sacred stream from which each cupful is to be preserved. But as with the Upanishads, the repetition, not only of ideas but of words, palls on the modern reader; and among alternative verses those must be chosen which seem most beautiful, most inspired. To regret the loss of continuity because of deletions is to miss the fact that the thought seems, from a modern viewpoint, to zigzag and cut across itself; and the same logical inconsistencies which appear in a page or two of the text appear massively in the whole. For the Bhagavad-Gita is a poem—in a way it is like a Homeric poem—its original elements arising in the minds of many poets. Unity is given—again as with the Homeric poems—not by a scribe or a commentator, but by a great artist, a supernal poetic craftsman, who felt the flow of the poem and, behind it, the flow of a mind and heart which were not, in the first instance, facing human tragedy logically, and which, in the second instance, needed to feel that the answers were logical because life was thereby less empty or less tragic. Indeed, as with Job and with Aeschylus, it can be magnificent *because* it is tragic.

In Western tragedy, however, suffering and death are transcended by being viewed as parts of an inscrutable cosmic whole, which for all its awfulness retains a majesty before which the knee bends; for the Bhagavad-Gita the tragedy is rejected, denied; not softened but excluded from the world of the real.

## YOGA

We come now to yoga, a system of reflection and self-control which applies Indian philosophy to individual salvation through the eradication of psychological error and the cultivation of pure self-knowledge. It has already been encountered in the Upanishads (page 62) and in the Bhagavad-Gita (page 94), but its systematic development by Patanjali constitutes one of the great psychological achievements of all time. Nothing is known about him for certain; the *Encyclopaedia Britannica* says that Patanjali is "an ancient Indian author or authors (!) perhaps of the second century of the Christian era."

Though not Sanskrit scholars ourselves, we shall first allow ourselves some notes on the Yoga Sutras of Patanjali. Patanjali has in mind the distinction between the self as knower, the process of knowing, the object known, and the investment in the object or cognitive contact with a known object. Such contact involves a certain contamination of the mental state which does the knowing. For in the Bhagavad-Gita, it is of the utmost importance for the self as knower to learn that it is *not* a part of the object known. This seems to be what is meant by undifferentiated consciousness. The self must learn that it is independent and utterly contentless as far as concerns matters known. The object known does, in fact, "contaminate" the knower (as field theory would require; see page 100). Aristotle said, "The mind is the thing known," and we suspect he meant what Patanjali means.

There is a continued insistence upon the purity—we would say the homogeneity—of the self, free of all fuzziness around the edges due to any sort of contacts with other things. This involves a true and absolute stability, unchangeability. This content is not a process, because a process would involve temporal change. The self is immune to all effects of change or time: there is a thinker over and above the temporal flow of thought process—a view attacked by William James in the celebrated chapter on "The Stream of Thought." James said that it is only a subsequent state of mind which can latch on to, or catch up to, or recall an earlier state of mind; for him there is no pure knower, a changeless and out-of-contact kind of soul such as is posited by Patanjali.

It is easy to see why, from Patanjali's viewpoint, there has to be a long period of discipline to keep the temporal flow in its place. That is, to recognize that it exists as a reality which can be known and discussed, *without* the soul's becoming involved in the false belief that the soul itself has this kind of temporal flowing quality. As we understand Patanjali, he is not saying that the flow of time is unreal. He is saying simply that it exists outside the sphere represented by the pure self. It takes the self, however, a great deal of time and much discipline to free itself from the contamination which would be involved in participation in this temporal flow. There are many exercises involving breath control, concentration on the tip of the nose, and so forth, until the self gradually recognizes its own independence of all these things.

The central idea appears to be the distinction between the changeless, essentially contentless supreme self and the mind-stuff or conscious process into which, largely through subliminal action, hindrances and bad karma are thrown. A fundamental training demand for the yogin is to perceive, and not merely understand,* the complete independence of the self from all mental content.

But the primary task of the yogin and of Patanjali himself is to learn to see, learn to know, directly apprehend with certainty the reality and changelessness of the self. From a modern Western viewpoint, the possibility that thought is guided by suggestion—so that you mistake for reality what you have learned to view as reality—appears not to be recognized at all. There is, consequently, no way in which the yogin can perceive that he is bringing upon himself the perceptual interpretation which his training system presupposes. The yogin learns to see and to apprehend exactly as

---

* Note the difference, in psychoanalysis, between "intellectualizing" or sheer cognitive recognition of a situation and the true "working through" of the situation.

the devotees of many other profoundly dogmatic religious systems do. The motivation of the yogin is apparently the same as that of Descartes, who would be horrified to discover that he did not exist as an entity, but only as a stream of thought.

We are disturbed by the way Patanjali handles the affective life. It seems to us that there is a constant emphasis upon passionlessness, which means, upon close analysis, the lack of all positive affect. At times pain, in the sense of unpleasantness or "negative affect," seems to be the only affect considered. At times the teaching seems to favor getting free of all affect. There are, nevertheless, references to joy, in a context which seems to imply that the yogin has joy or is pursuing joy as part of his training. This may be the usual confusion which we have also in the Western tradition, in which we repudiate "pleasure" but look for a "higher pleasure." In this sense it is said: "What constitutes the pleasure of love in this world and what the supreme pleasure of heaven are both not to be compared with the sixteenth part of the pleasure of dwindled craving." This could still mean, with Socrates, that pleasure is the absence of pain; but there is much more than deliverance from pain here, namely, a very intense continuing pleasure. In all this linguistic confusion, it seems to us that Patanjali is saying, as many in the West have done, that peace or serenity is both affectless and also joyful; and we can make sense out of this only if life for Patanjali is axiomatically regarded as painful. It would be interesting to know whether he differed sharply from the Buddhists in these matters (as he does differ with them in regard to the question whether the self is an absolute timeless entity rather than a flow of thought).

But aside from questions of logical consistency, note the deep feeling that this life is not good. Note the extraordinary irreversible character of the acts of escape: "This unwavering discriminative discernment is the means of escape. After this, erroneous perception tends to become reduced to the condition of burned seed. And its failure to reproduce itself is the path to release, the way of approach to escape." Such language as that of escape shows how profoundly the yogin repudiates the world of this life. Note the frequent recurrence of ethical issues: "Therefore let the yogin con sider first what is good for all creatures, and then speak with abstinence from falsehood."

The yogin gets along without external stimulation and also without memory, and achieves the "self's isolation." "The self's energy of thought becomes isolated, since it is grounded in itself and is not related to the *sattva* of the thinking substance. Its continuance thus forevermore is isolation."

The selections that follow, from J. H. Woods' translation of the source material, are in themselves a commentary; the actual text of the Yoga Sutras is shown in boldface type.*

### Book I. Concentration or Samadi

**1. Now the exposition of yoga [is to be made].**
The expression "now" indicates that a distinct topic commences here. The authoritative book which expounds yoga is to be understood as commenced. [To give a provisional definition:] yoga is concentration; but this is a quality of the mind-stuff (*citta*) which belongs to all the stages. The stages of the mind-stuff are these: the restless (*ksipta*), the infatuated (*mudha*), the distracted (*viksipta*), the single-in-intent (*ekagra*), and the restricted (*niruddha*). Of these [stages the first two have nothing to do with yoga and even] in the distracted state of the mind [its] concentration is [at times] overpowered by [opposite] distractions and [consequently] it cannot properly be called yoga. But that [state] which, when the mind is single-in-intent, fully illumines a distinct and real object and causes the hindrances (*kleca*) to dwindle, slackens the bonds of karma, and sets before it as a goal the restriction [of all fluctuations], is called the yoga in which there is consciousness of an object (*samprajnata*). This [conscious yoga], however, is accompanied by deliberation [upon coarse objects], by reflection [upon subtle objects], by joy, by the feeling-of-personality (*asmita*). This we shall set forth later. But when there is restriction of all the fluctuations (*vrtti*) [of the mind-stuff], there is the concentration in which there is no consciousness [of an object]. The intent of the following sutra is to state the distinguishing characteristic of this [yoga].

**2. Yoga is the restriction of the fluctuations of mind-stuff.**
By the non-use of the word "all" [before "the fluctuations"], [the yoga which is] conscious [of objects] is also included under the denomination of yoga. Now mind-stuff has three aspects (*guna*), as appears from the

* From J. H. Woods (trans.), *The Yoga-System of Patanjali* (Cambridge, Mass.: Harvard University Press, 1914).

fact that it has a disposition to vividness (*prakhya*), to activity (*pravrtti*), and to inertia (*sthiti*). For the mind-stuff's [aspect] *sattva*, which is vividness, when commingled with *rajas* and *tamas*, acquires a fondness for supremacy and for objects-of-sense; while the very same [constituent-aspect, *sattva*,] when pervaded with *tamas*, tends toward demerit and non-perception and passionateness and toward a failure of [its own rightful] supremacy; [and] the very same [*sattva*]—when the covering of error has dwindled away—illumined now in its totality (*sarvatas*), but faintly pervaded by *rajas*, tends toward merit and knowledge and passionlessness and [its own rightful] supremacy; [and] the very same [*sattva*]—the stains of the last vestige of *rajas* once removed—grounded in itself and being nothing but the discernment (*khyati*) of the difference between the *sattva* and the Self (*purusha*), tends toward the Contemplation of the Rain-cloud of [knowable] Things. The designation given by contemplators (*dhyayin*) to this [kind of mind-stuff] is the highest Elevation (*prasamkhyana*). For the Energy of Intellect (*citi-cakti*) is immutable and does not unite [with objects]; it has objects shown to it and is undefiled [by constituent-aspects] and is unending. Whereas this discriminate discernment (*viveka-khyati*), whose essence is *sattva*, is [therefore] contrary to this [Energy of Intellect and is therefore to be rejected]. Hence the mind-stuff being disgusted with this [discriminative discernment] restricts even this Insight. When it has reached this state, [the mind-stuff], [after the restriction of the fluctuations,] passes over to subliminal impressions (*samskara*). This is the [so-called] seedless concentration. In this state nothing becomes an object of consciousness: such is concentration not conscious [of objects]. Accordingly the yoga [which we have defined as] the restriction of the fluctuations of the mind-stuff is twofold.

* * *

The mind being in this [unconscious] state, what will then be the condition of the Self? For it is the essence (*atman*) [of the Self to receive] knowledge (*bodha*) [reflected upon it] by the thinking-substance (*buddhi*), [as this in its turn receives the impression of external objects, and in this case] there is a [total] absence of objects [in the thinking-substance].

### 3. Then the Seer [that is, the Self,] abides in himself.
At that time the Energy of Intellect is grounded in its own self, as [it is] when in the state of Isolation. But when the mind-stuff is in its emergent state, [the Energy of Intellect], although really the same, [does] not [seem] so.

* * *

**4. At other times it [the Self] takes the same form as the fluctuations [of mind-stuff].**

In the emergent state [of the subliminal-impressions], the Self has fluctuations which are not distinguished from fluctuations of the mind-stuff; and so we have a sutra [of Pancacikha], "There is only one appearance [for both]—that appearance is knowledge." The mind-stuff is like a magnet; and, as an object suitable to be seen [by the Self as Witness], it gives its aid [to the Self] by the mere fact of being near it, and thus the relation between it and the Self is that between property (*svam*) and proprietor (*svamin*). Hence the reason why the Self experiences (*bodha*) the fluctuations of the mind-stuff is its beginning-less correlation [with the thinking-substance].

.    .    .

**5. The fluctuations are of five kinds and are hindered or unhindered.**

The hindered (*klista*) are those which are caused by the hindrances (*kleca*) [undifferentiated-consciousness, etc.] and are the field for growth of the accumulation of the latent-deposits of karma; the un-hindered have discriminative discernment as their object and thus obstruct the task (*adhikara*) of the aspects (*guna*). These are still un-hindered even when they occur in the stream of the hindered. For even in the midst of the hindered [fluctuations] they are unhindered; while in the midst of the unhindered [they are] hindered. Corresponding sub-liminal-impressions are produced by nought else than [these] fluctuations, and fluctuations [are made] by subliminal-impressions. In this wise, the wheel of fluctuations and subliminal-impressions ceaselessly rolls on [until the highest concentration is attained]. Operating in this wise, this mind-stuff, having finished its task, abides in its own likeness, or [rather] becomes resolved [into primary substance].—These, either hindered or unhindered, are the fivefold fluctuations.

.    .    .

**6. Sources-of-valid-ideas and misconceptions and predicate relations and sleep and memory.**

.    .    .

**7. Sources-of-valid-ideas are perception and inference and verbal-communication.**

Perception is that source-of-valid-ideas [which arises as a modification of the inner-organ] when the mind-stuff has been affected by some external thing through the channel of the sense-organs. This fluctuation is directly related to that [object], but, whereas the intended-object (*artha*) consists of a genus and of a particular, it [the fluctuation] is chiefly concerned with the ascertainment of the particular [the genus being subordinate in perception to the particular]. The result [of perception] is an illumination by the Self (*pauruseya*) of a fluctuation which belongs to the mind-stuff, [an illumination which is] undistinguished (*a-vicista*), [that is, one in which the Self does not distinguish itself from the thinking-substance]. ii. Inference is [that] fluctuation [of the mind-stuff] which refers (-*visaya*) to that (*tat-*) relation (*sambandha*) which is present in things belonging to the same class as the subject-of-the-illation (*anumeya*) and absent from things belonging to classes different [from that of the subject-of-the-illation]; and it is chiefly concerned with the ascertainment of the genus. Thus, for instance, the moon and stars possess motion, because, like [any man, for instance,] Chaitra, they get from one place to another; and because [negatively] the Vindhya [mountain-range] does not get [from one place to another, it] does not possess motion. iii. A thing which has been seen or inferred by a trustworthy person is mentioned by word in order that his knowledge [thereof] may pass over to some other person. The fluctuation [in the mind-stuff] of the hearer which arises from that word and which relates to the object-intended by that [word] (*tad-artha-visaya*) is a verbal-communication. That verbal-communication is said to waver, the utterer of which declares an incredible thing, not a thing which he himself has seen or inferred; but if the original utterer has himself seen or inferred the thing, [then the verbal-communication] would be unwavering.

.        .        .

**8. Misconception is an erroneous idea (jnana) not based on that form [in respect of which the misconception is entertained].**
Why is it not a source-of-a-valid-idea? Because it is inhibited by the source-of-a-valid-idea, for the reason that the source-of-a-valid-idea has as its object a positive fact. In such cases there is evidently an inhibition of the source-of-the-invalid-idea by the source-of-the-valid-idea, as for instance the [erroneous] visual-perception of two moons is inhibited by the actual (*sad-visaya*) visual-perception of one moon. This [fluctuation, namely, misconception] proves to be that [well-known] five-jointed undifferentiated-consciousness. These same [are known] by peculiar technical designations: Obscurity and Infatuation and Extreme Infatuation and Darkness and Blind-Darkness.

.        .        .

**12. The restriction of them is by [means of] practice and passionlessness.**
The so-called river of mind-stuff, whose flow is in both directions, flows toward good and flows toward evil. Now when it is borne onward to

Isolation [*kaivalya*], downward toward discrimination, then it is flowing unto good; when it is borne onward to the whirlpool-of-existence, downward toward non-discrimination, then it is flowing unto evil. In these cases the stream toward objects is dammed by passionlessness, and the stream toward discrimination has its flood-gate opened by practice in discriminatory knowledge. Thus it appears that the restriction of the mind-stuff is dependent [for its accomplishment upon means] of both kinds, [practice and passionlessness].

.    .    .

**13. Practice (abhyasa) is [repeated] exertion to the end that [the mind-stuff] shall have permanence in this [restricted state].**
Permanence is the condition of the unfluctuating mind-stuff when it flows on in undisturbed calm. Practice is an effort (*prayatna*) with this end in view—a [consequent] energy, a persevering struggle—the pursuit (*anusthana*) of the course-of-action-requisite thereto with a desire of effectuating this [permanence].

.    .    .

**14. But this [practice] becomes confirmed when it has been cultivated for a long time and uninterruptedly and with earnest attention.**
[Practice,] when it has been cultivated for a long time, cultivated without interruption, and carried out with self-castigation and with continence and with knowledge and with faith—in a word, with earnest attention—becomes confirmed. In other words it is not likely to have its object suddenly overpowered by an emergent subliminal-impression.

.    .    .

**15. Passionlessness is the consciousness of being master on the part of one who has rid himself of thirst for either seen or revealed objects.**
The mind-stuff (*citta*)—if it be rid of thirst for objects that are seen, such as women, or food and drink, or power—if it be rid of thirst for the objects revealed [in the Vedas], such as the attainment of heaven or of the discarnate state or of resolution into primary matter—if, even when in contact with objects either super-normal or not, it be, by virtue of Elevation (*prasamkhyana*), aware of the inadequateness of objects—[then the mind-stuff] will have a consciousness of being master, [a consciousness] which is essentially the absence of immediate-experience (*abhoga*) [and] has nothing to be rejected or received, [and that consciousness is] passionlessness.

.    .    .

**16. This [passionlessness] is highest when discernment of the Self results in thirstlessness for qualities [and not merely for objects].**

[One yogin becomes] passionless on knowing the inadequateness of [all] objects, seen or revealed. Through practice in the vision of the Self, [another yogin,] because his thinking-substance is satiated with a perfect discrimination, resulting from the purity of this [vision], [between the qualities (*guna*) and the Self], [becomes] passionless with regard to [all] qualities whether perceptible or not perceptible. Thus passionlessness is of two kinds. Of these [two], the latter is nothing but an undisturbed calm of perception [untouched by any objects whatsoever]. And at the rising of this [state, the yogin] on whom this insight has dawned, thus reflects within himself, "That which was to be attained (*prapaniya*) has been attained; the hindrances which should have dwindled have dwindled . . ."

.     .     .

**17. [Concentration becomes] conscious [of its object] by assuming forms either of deliberation [upon coarse objects] or of reflection [upon subtile objects] or of joy or of the sense-of-personality.**

Deliberation (*vitarka*) is the mind-stuff's coarse direct-experience (*abhoga*) when directed to its supporting [object]. Reflection (*vieara*) is the subtile [direct-experience]. Joy is happiness. The sense-of-personality is a feeling (*samvid*) which pertains to one self [wherein the Self and the personality are one]. Of these [four] the first, [that is, deliberation] which has [all] the four associated together is concentration deliberating [upon coarse objects]. The second, [that is, reflection,] which has deliberation subtracted [from it] is [concentration] reflecting [upon subtile objects]. The third, [that is, joy,] which has reflection subtracted from it, is [concentration] with [the feeling] of joy. The fourth, [that is, the sense-of-personality,] which has this [joy] subtracted from it, is [concentration] which is the sense-of-personality and nothing more. All these kinds of concentrations have an object upon which they rest.

.     .     .

**18. The other [concentration which is not conscious of objects] consists of subliminal-impressions only [after objects have merged], and follows upon that practice which effects the cessation [of fluctuations].**

The concentration which is not conscious [of objects] is that restriction of the mind-stuff in which only subliminal-impressions are left and in which all fluctuations have come to rest. The higher passionlessness is a means for effecting this. For practice when directed toward any supporting-object is not capable of serving as an instrument to this [concentration not conscious of an object]. So the supporting-object [for this concentration] is [the Rain-cloud of knowable things] which effects this cessation [of fluctuations] and has no [perceptible] object. For (*ca*) [in this concentration] there is no object-intended. Mind-stuff, when engaged in the practice of this [imperceptible object], seems as if it were itself nonexistent and without any supporting-object. Thus [arises] that concentration [called] seedless, [without sensational stimulus], which is not conscious of objects.

.    .    .

**20. [Concentration not conscious of objects] which follows upon belief [and] energy [and] mindfulness [and] concentration [and] insight, is that to which the others [the yogins] attain.**
[That concentration not conscious of objects, which is] caused by [spiritual] means is that to which yogins attain. Belief is the mental approval [of concentration]; for, like a good mother, it protects the yogin. For him [thus] believing and setting discrimination [before him] as his goal there is the further (*upa*) attainment of energy. For him who has reached the further attainment of energy mindfulness is at hand. And when mindfulness is at hand the mind-stuff is self-possessed and becomes concentrated. When his mind-stuff has become concentrated he gains as his portion the discrimination of insight, by which he perceives things as they really are. Through the practice of these means and through passionlessness directed to this end there [finally] arises that concentration which is not conscious [of any object].
Now, by the yogin who has recognized the power of the word to express the thing,

.    .    .

**28. Repetition of it and reflection upon its meaning [should be made].**
The repetition of the Mystic Syllable, and reflection upon the Icvara who is signified by the Mystic Syllable. Then in the case of this yogin who thus repeats the Mystic Syllable and reflects upon its meaning, mind-stuff attains to singleness-of-intent. And so it hath been said,

Through study let him practice yoga;
Through yoga let him meditate on study.
By perfectness in study and in yoga
Supreme Soul shines forth, clearly.

.    .    .

105

**30. Sickness and languor and doubt and heedlessness and listlessness and worldliness (avirati) and erroneous perception and failure to attain any stage [of concentration] and instability in the state [when attained] —these distractions of the mind-stuff are the obstacles.**

There are nine obstacles, the distractions of the mind-stuff. These appear together with the fluctuations of the mind-stuff. And they are not found where the aforesaid fluctuations of mind-stuff are not. Sickness is a disorder in the humors [of the body] or in the secretions or in the organs. Languor is a lack of activity in the mind-stuff. Doubt is a kind of thinking which touches both alternatives [of a dilemma], so that one thinks "This might be so; might not be so." Heedlessness is a lack of reflection upon the means of attaining concentration. Listlessness is a lack of effort due to heaviness of body or of mind-stuff. Worldliness is greed of the mind-stuff; and its essence lies in addiction to objects of sense. Erroneous perception is the thinking of misconceptions. Failure to attain any stage is not attaining any stage of concentration. Instability in the state [when attained] is the failure of the mind-stuff to remain in the stage attained. If the concentrated stage of development had been reached, [the mind-stuff] would, of course, have remained in it.—Thus it is that these distractions are called the nine blemishes of yoga [and] the nine foes of yoga [and] the obstacles of yoga.

.    .    .

**31. Pain and despondency and unsteadiness of the body and inspiration and expiration are the accompaniments of the distractions.**

Pain proceeding from self [and] pain proceeding from living creatures and pain proceeding from the gods. Pain is that by which living beings are stricken down and for the destruction of which they struggle. Despondency is agitation of mind due to an impediment [to the fulfillment] of a desire. Unsteadiness of the body is that which makes it unsteady [and] makes it tremble. Inspiration is breathing which sips in the air which is outside. Expiration is that which makes abdominal air flow outward. These are the accompaniments of the distractions. These occur in one whose mind-stuff is distracted. These do not occur in one whose mind-stuff is concentrated.

.    .    .

**32. To check them [let there be] practice upon a single entity.**

To check them let [the yogin] practice his mind-stuff by making it rest upon a single entity. But one whose mind-stuff is nothing more than an idea limited to one object after another, and is momentary (*ksanika*)— of this [Buddhist] the mind-stuff as a whole is surely not single-in-intent and it is surely not distracted. But if this [mind-stuff when single-in-intent] is withdrawn from all [objects] and concentrated upon one [entity], then it may be said to be single-in-intent [and] hence not limited to one object after another. If, on the other hand—[in the opinion] of him who maintains that the mind-stuff becomes single-in-intent as a stream of similar ideas—singleness-in-intent be a property of the mind-stuff [conceived] as a stream, then the mind-stuff [conceived as] a stream could not be a single thing, because [as he insists] it changes from moment to moment. If however [it be maintained that] singleness-of-intent is a property of an idea only in so far as it forms a part of the stream, then—whether it consist in a stream of similar ideas or in a stream of dissimilar ideas—it is all of it in nowise other than single-in-intent, inasmuch as it is limited to one object after another, and the fact that mind-stuff is distracted is unexplained. Therefore it may be said that mind-stuff is a single thing [and] has many intended objects [and] is stable.

### Book IV. Isolation or Kaivalya

**22. The Intellect (citi) which unites not [with objects] is conscious of its own thinking-substance when [the mind-stuff] takes the form of that [thinking-substance by reflecting it].**

"For, the Energy of the experiencer, which is immutable and which unites not with objects, seeming to unite with something mutable [the thinking-substance], takes the form of the fluctuations which that [thinking-substance] undergoes. And [this Energy] being identified with a fluctuation of the thinking-substance in so far as it is nothing but an imitation of a fluctuation of the thinking-substance that has come under the influence of the intelligence (*caitanya*), it is termed a fluctuation of the thinking-substance." And in this sense it has been said, "That hiding-place in which the everlasting Brahman lies concealed is not an underworld nor mountain-chasm nor the darkness nor caverns of the sea, but is the working of the mind when not distinguished [from Himself]. So the sages tell."

Yoga is clearly a system of psychology in which the individual self—serene and changeless—is sharply separated from the thought processes going on within. But it is also a practical discipline, a form of schooling, in which a master transmits a skill of self-training to his pupils. Most books on yoga are concerned with such discipline. The great tradition has, moreover, been diversified or enriched by the development of various kinds of discipline. The tradition of Patanjali himself is maintained in the higher spiritual discipline of *Raja-yoga* (royal yoga). Notable is the physiological discipline of *Hatha-yoga*, the "yoga of force," which is concerned with the cultivation of extraordinary bodily control, and leads, it is claimed, to the capacity to arrest the heartbeat and the process of breathing. (From a Western viewpoint, the claims are still doubted, though active research by Western methods is in progress.)

## BUDDHISM

During the centuries of thought dominated by the Upanishads, many differences of opinion arose, often leading to new philosophic schools. We should not ordinarily call such schools new religions. But in the case of Prince Gautama, Lord Buddha, the intensity of his message, the earnestness of his disciples, and—within two centuries of his death—the development of a new systematic theology centering in his ideas, leave no doubt that a major religious philosophy and religious way of life had arisen.

As a boy and a young man, he had been protected from the world's suffering. But in his late twenties he encountered men undergoing suffering, disease, and death, and grasped that all the joys of life must end in tragedy. Leaving his wife and infant son one morning at daybreak, he went to seek a way of escape. After several years of searching, the Enlightenment came to him "under the bo tree" in Bihar (in northeastern India). After repeated messages to his disciples, he began to give formal sermons, the text of which is thought to be fairly well preserved.

His messages had enormous vitality for the hundreds of millions of Indians, Tibetans, Ceylonese, Burmese, Thais, Chinese, Koreans, Japanese, and others to whom it came. We do not offer Buddhism here as a "great psychological system." But it emphasizes (1) renunciation of the desires which lead to frustration and suffering; (2) a "noble eightfold path" leading very practically to escape from desire; (3) a loss of the personal identity that is rooted in desires; (4) an ultimate absorption after many rebirths into a blissful state of nirvana or serenity in which selfhood is lost.

Four respects in which the psychology is new are (1) the

emphasis on practical action as a source of escape; (2) the emphasis on benevolence and compassion, both for their direct appeal and for their utility in lifting one out of the cycle of rebirth; (3) the creation of a spirit of brotherhood available both to the priest and to the common people; (4) the denial of any central and persisting soul or atman. The flux of experience goes on, but there is no abiding core of changeless selfhood.

Prince Gautama came to see that there is a genuine path of escape from the sufferings of this life. He began his career as a teacher with a sermon at Benares. Equally futile, he says, are the pursuit of pleasure and the empty process of self-mortification. The cause of pain can be eradicated through the noble eightfold path: right view, right thought, right speech, right action, right livelihood, right effort, right mindfulness, right concentration. The cure for sufferings of this world, as with the Hindu, lies in psychology— that is, in the understanding of mind, heart, and will. But instead of the formalism of the great Hindu priesthood, something very warm, simple, and direct is offered, couched in terms of simple ways of dealing with the environment, as well as with the inner world, in a message intelligible to all. The Sermon at Benares, which we first quote here, is a part of the earliest canonical Buddhist literature included in the sacred writings gathered and accepted by the great Buddhist congresses of two centuries after his death. Our second selection, the Brahmana, is from the Dhammapada, a very ancient Buddhist text, parts of which are thought possibly to be from Buddha himself.

## The Sermon at Benares*

On seeing their old teacher approach, the five *bhikkhus* agreed among themselves not to salute him, nor to address him as a master, but by his name only. "For," so they said, "he has broken his vow and has abandoned holiness. He is no *bhikkhu* but Gautama, and Gautama has become a man who lives in abundance and indulges in the pleasures of worldliness."

But when the Blessed One approached in a dignified manner, they involuntarily rose from their seats and greeted him in spite of their

* Trans. by Samuel Beal, from Paul Carus (ed.), *The Gospel of Buddha* (La Salle, Ill.: The Open Court Publishing Company, 1915).

resolution. Still they called him by his name and addressed him as "friend Gautama."

When they had thus received the Blessed One, he said: "Do not call the Tathagata by his name nor address him as 'friend,' for he is the Buddha, the Holy One. The Buddha looks with a kind heart equally on all living beings, and they therefore call him 'Father.' To disrespect a father is wrong; to despise him, is wicked.

"The Tathagata," the Buddha continued, "does not seek salvation in austerities, but neither does he for that reason indulge in worldly pleasures, nor live in abundance. The Tathagata has found the middle path.

"There are two extremes, O *bhikkhus*, which the man who has given up the world ought not to follow—the habitual practice, on the one hand, of self-indulgence which is unworthy, vain, and fit only for the worldly-minded—and the habitual practice, on the other hand, of self-mortification, which is painful, useless and unprofitable.

"Neither abstinence from fish or flesh, nor going naked, nor shaving the head, nor wearing matted hair, nor dressing in a rough garment, nor covering oneself with dirt, nor sacrificing to Agni, will cleanse a man who is not free from delusions.

"Reading the Vedas, making offerings to priests, or sacrifices to the gods, self-mortification by heat or cold, and many such penances performed for the sake of immortality, these do not cleanse the man who is not free from delusions.

"Anger, drunkenness, obstinacy, bigotry, deception, envy, self-praise, disparaging others, superciliousness and evil intentions constitute uncleanness; not verily the eating of flesh.

"A middle path, O *bhikkhus*, avoiding the two extremes, has been discovered by the Tathagata—a path which opens the eyes, and bestows understanding, which leads to peace of mind, to the higher wisdom, to full enlightenment, to nirvana!

"What is that middle path, O *bhikkhus*, avoiding these two extremes, discovered by the Tathagata—that path which opens the eyes, and bestows understanding, which leads to peace of mind, to the higher wisdom, to full enlightenment, to nirvana?

"Let me teach you, O *bhikkhus*, the middle path, which keeps aloof from both extremes. By suffering, the emaciated devotee produces confusion and sickly thoughts in his mind. Mortification is not conducive even to worldly knowledge; how much less to a triumph over the senses!

"He who fills his lamp with water will not dispel the darkness, and he who tries to light a fire with rotten wood will fail. And how can any one be free from self by leading a wretched life, if he does not succeed in quenching the fires of lust, if he still hankers after either worldly or heavenly pleasures? But he in whom self has become extinct is free from lust; he will desire neither worldly nor heavenly pleasures, and the satisfaction of his natural wants will not defile him. However, let him be moderate, let him eat and drink according to the needs of the body.

"Sensuality is enervating; the self-indulgent man is a slave to his passions, and pleasure-seeking is degrading and vulgar.

"But to satisfy the necessities of life is not evil. To keep the body in good health is a duty, for otherwise we shall not be able to trim the lamp of wisdom, and keep our mind strong and clear. Water surrounds the lotus-flower, but does not wet its petals.

"This is the middle path, O *bhikkhus*, that keeps aloof from both extremes."

And the Blessed One spoke kindly to his disciples, pitying them for their errors, and pointing out the uselessness of their endeavors, and the ice of ill-will that chilled their hearts melted away under the gentle warmth of the Master's persuasion.

Now the Blessed One set the wheel of the most excellent law rolling, and he began to preach to the five *bhikkhus*, opening to them the gate of immortality, and showing them the bliss of nirvana.

The Buddha said:

"The spokes of the wheel are the rules of pure conduct: justice is the uniformity of their length; wisdom is the tire; modesty and thoughtfulness are the hub in which the immovable axle of truth is fixed.

"He who recognizes the existence of suffering, its cause, its remedy, and its cessation has fathomed the four noble truths. He will walk in the right path.

"Right views will be the torch to light his way. Right aspirations will be his guide. Right speech will be his dwelling place on the road. His gait will be straight, for it is right behavior. His refreshments will be the right way of earning his livelihood. Right efforts will be his steps: right thoughts his breath; and right contemplation will give him the peace that follows in his footprints.

"Now, this, O *bhikkhus*, is the noble truth concerning suffering:

"Birth is attended with pain, decay is painful, disease is painful, death is painful. Union with the unpleasant is painful, painful is separation from the pleasant; and any craving that is unsatisfied, that too is painful. In brief, bodily conditions which spring from attachment are painful.

"This, then, O *bhikkhus*, is the noble truth concerning suffering.

"Now this, O *bhikkhus*, is the noble truth concerning the origin of suffering:

"Verily, it is that craving which causes the renewal of existence, accompanied by sensual delight, seeking satisfaction now here, now there, the craving for the gratification of the passions, the craving for a future life, and the craving for happiness in this life.

"This, then, O *bhikkhus*, is the noble truth concerning the origin of suffering.

"Now this, O *bhikkhus*, is the noble truth concerning the destruction of suffering:

"Verily, it is the destruction, in which no passion remains, of this very thirst; it is the laying aside of, the being free from, the dwelling no longer upon this thirst.

"This, then, O *bhikkhus*, is the noble truth concerning the destruction of suffering.

"Now this, O *bhikkhus*, is the noble truth concerning the way which leads to the destruction of sorrow. Verily! it is this noble eightfold path; that is to say:

"Right views; right aspirations; right speech; right behavior; right livelihood; right effort; right thoughts; and right contemplation.

"This, then, O *bhikkhus*, is the noble truth concerning the destruction of sorrow.

"By the practice of lovingkindness I have attained liberation of heart, and thus I am assured that I shall never return in renewed births. I have even now attained nirvana."

And when the Blessed One had thus set the royal chariot wheel of truth rolling onward, a rapture thrilled through all the universes.

The *devas* left their heavenly abodes to listen to the sweetness of the truth; the saints that had parted from life crowded around the great teacher to receive the glad tidings; even the animals of the earth felt the bliss that rested upon the words of the Tathagata: and all the creatures of the host of sentient beings, gods, men, and beasts, hearing the message of deliverance, received and understood it in their own language.

And when the doctrine was propounded, the venerable Kondanna, the oldest one among the five *bhikkhus*, discerned the truth with his mental eye, and he said: "Truly, O Buddha, our Lord, thou hast found the truth!" Then the other *bhikkhus* too, joined him and exclaimed: "Truly, thou art the Buddha, thou hast found the truth."

And the *devas* and saints and all the good spirits of the departed generations that had listened to the sermon of the Tathagata, joyfully received the doctrine and shouted: "Truly, the Blessed One has founded the kingdom of righteousness. The Blessed One has moved the earth; he has set the wheel of Truth rolling, which by no one in the universe, be he god or man, can ever be turned back. The kingdom of Truth will be preached upon earth; it will spread; and righteousness, good-will, and peace will reign among mankind."

### The Brahmana*

Stop the stream valiantly, drive away the desires, O Brahmana! When you have understood the destruction of all that was made, you will understand that which was not made.

If the Brahmana has reached the other shore in both laws, in restraint and contemplation, all bonds vanish from him who has obtained knowledge.

He for whom there is neither the hither nor the further shore, nor both, him, the fearless and unshackled, I call indeed a Brahmana.

He who is thoughtful, blameless, settled, dutiful, without passions, and who has attained the highest end, him I call indeed a Brahmana.

The sun is bright by day, the moon shines by night, the warrior is bright in his armor, the Brahmana is bright in his meditation; but Buddha, the Awakened, is bright with splendor day and night.

Because a man is rid of evil, therefore he is called Brahmana; because he walks quietly, therefore he is called Samana; because he has sent away his own impurities, therefore he is called Pravragita (Pabbagita, a pilgrim).

No one should attack a Brahmana, but no Brahmana, if attacked, should let himself fly at his aggressor! Woe to him who strikes a Brahmana, more woe to him who flies at his aggressor!

It advantages a Brahmana not a little if he holds his mind back from the pleasures of life; the more all wish to injure has vanished, the more all pain will cease.

Him I call indeed a Brahmana who does not offend by body, word, or thought, and is controlled on these three points.

He from whom he may learn the law, as taught by the Well-awakened (Buddha), him let him worship assiduously, as the Brahmana worships the sacrificial fire.

A man does not become a Brahmana by his plaited hair, by his family, or by birth; in whom there is truth and righteousness, he is blessed, he is a Brahmana.

What is the use of plaited hair, O fool! what of the raiment of goatskins? Within thee there is ravening, but the outside thou makest clean.

The man who wears dirty raiments, who is emaciated and covered with veins, who meditates alone in the forest, him I call indeed a Brahmana.

I do not call a man a Brahmana because of his origin or of his mother.

* Trans. by F. M. Müller, from Lin Yutang (ed.), *The Wisdom of India and China* (New York: Random House, 1942). Copyright 1942 by Random House, Inc. Reprinted by permission.

He is indeed arrogant, and he is wealthy: but the poor, who is free from all attachments, him I call indeed a Brahmana.

Him I call indeed a Brahmana who, after cutting all fetters, never trembles, is free from bonds and unshackled.

Him I call indeed a Brahmana who, after cutting the strap and the thong, the rope with all that pertains to it, has destroyed all obstacles, and is awakened.

Him I call indeed a Brahmana who, though he has committed no offense, endures reproach, stripes, and bonds: who has endurance for his force, and strength for his army.

Him I call indeed a Brahmana who is free from anger, dutiful, virtuous, without appetites, who is subdued, and has received his last body.

Him I call indeed a Brahmana who does not cling to sensual pleasures, like water on a lotus leaf, like a mustard seed on the point of a needle.

Him I call indeed a Brahmana who, even here, knows the end of his own suffering, has put down his burden, and is unshackled.

Him I call indeed a Brahmana whose knowledge is deep, who possesses wisdom, who knows the right way and the wrong, and has attained the highest end.

It is not possible to read much Buddhist literature and commentary upon it without encountering two related questions: (1) What *is* the state of nirvana to which all roads seem to lead, the state of blessed "annihilation of desire"—or should we say "annihilation of individuality"? (2) How do the varying answers to the question relate to the great schism in Buddhism, between the "lesser vehicle," or *hinayana*, and the "greater vehicle," or *mahayana*?

We cannot afford to get entangled in theology, and shall ruthlessly limit ourselves to two issues: (1) The struggle to get rid of individuality, which we have seen throughout the development of Hinduism, took in Buddhism the form of inquiry into what happens when the stream of ideas, associations, feeling, and mental processes generally, cease at death. Since Buddhism cannot accept an independent soul surviving death, and yet must recognize the universal continuity of life, it compares the individual's thoughts to a candle-flame which can ignite one candle after another though the wick of each is burned up. (2) Another figure of speech:

Paradoxical though it may seem: There is a path to walk on, there is walking being done, but there is no traveler. There are deeds being done, but there is no doer. There is a blowing of the air, but there is no wind that does the blowing. The thought of self is an error and all existences are as hollow as the plantain tree and as empty as twirling water bubbles.

Therefore as there is no self, there is no transmigration of a self; but there are deeds and the continued effect of deeds. There is a rebirth of karma; there is reincarnation. This rebirth, this reincarnation, this reappearance of the conformations is continuous and depends on the law of cause and effect.*

The karma, the psychological continuity from moment to moment, continues. And it will continue until desire, which keeps us going, comes to an end. The blissful end of the cycle of rebirths is absorption into cosmic reality—nirvana—in which individuality is lost.

But in Mahayana Buddhism some great souls, nearing the fulfillment of these tremendous aspirations, choose to delay the final passage to nirvana, to aid other souls on their way. They stop at the brink—as if hearing the words "absent thee from felicity awhile"; or to use a humbler figure of speech, they are like "rovers" in croquet, who do not "go out" by hitting the last stake, but keep in the game to be helpful. The fact of this choice, made by the Bodhisattva, who has taken all but the last step to nirvana, becomes of huge importance in the developing emphasis upon compassion, mercy, and utter selflessness, so central in the ethics of Buddhism. The following passage is from the Surangama, a Chinese Buddhist text.

### What Is Nirvana?**

Then said Mahamati to the Blessed One: Pray tell us about nirvana?

The Blessed One replied: The term "nirvana" is used with many different meanings, by different people, but these people may be divided into four groups: There are people who are suffering, or who are afraid

---

* From *Sayings of Buddha* (Mount Vernon, N.Y.: Peter Pauper Press, 1957).

* * Trans. by Wai-Tao and Dwight Goddard, from Dwight Goddard (ed.), *A Buddhist Bible* (Thetford, Vt., 1938). Copyright 1938 by E. P. Dutton & Co., Inc. Reprinted by permission of the publishers.

of suffering, and who think of nirvana; there are the philosophers who try to discriminate nirvana; there are the class of disciples who think of nirvana in relation to themselves; and, finally there is the nirvana of the Buddhas.

Those who are suffering or who fear suffering, think of nirvana as an escape and a recompense. They imagine that nirvana consists in the future annihilation of the senses and the sense-minds; they are not aware that universal mind and nirvana are one, and that this life-and-death world and nirvana are not to be separated. These ignorant ones, instead of meditating on the imagelessness of nirvana, talk of different ways of emancipation. Being ignorant of, or not understanding, the teachings of the Tathagatas, they cling to the notion of nirvana that is outside what is seen of the mind and, thus, go on rolling themselves along with the wheel of life and death.

As to nirvanas discriminated by the philosophers: there really are none. Some philosophers conceive nirvana to be found where the mind-system no more operates owing to the cessation of the elements that make up personality and its world; or is found where there is utter indifference to the objective world and its impermanency. Some conceive nirvana to be a state where there is no recollection of the past or present, just as when a lamp is extinguished, or when a seed is burned, or when a fire goes out; because then there is the cessation of all the substrate, which is explained by the philosophers as the non-rising of discrimination. But this is not nirvana, because nirvana does not consist in simple annihilation and vacuity.

Again, some philosophers explain deliverance as though it were the mere stopping of discrimination, as when the wind stops blowing, or as when one by self-effort gets rids of the dualistic view of knower and known, or gets rid of the notions of permanency and impermanency; or gets rid of the notions of good and evil; or overcomes passion by means of knowledge—to them nirvana is deliverance. Some, seeing in "form" the bearer of pain, are alarmed by the notion of "form" and look for happiness in a world of "no-form." Some conceive that in consideration of individuality and generality recognizable in all things inner and outer, that there is no destruction and that all beings maintain their being forever and, in this eternality, see nirvana. Others see the eternality of things in the conception of nirvana as the absorption of the finite-soul in Supreme Atman; or who see all things as a manifestation of the vital-force of some Supreme Spirit to which all return; and some, who are especially silly, declare that there are two primary things, a primary substance and a primary soul, that react differently upon each other and thus produce all things from the transformations of qualities; some think that the world is born of action and interaction and that no other cause is necessary; others think that Ishvara is the free creator of all things; clinging to these foolish notions, there is no awakening, and they consider nirvana to consist in the fact that there is no awakening.

Some imagine that nirvana is where self-nature exists in its own right, unhampered by other self-natures, as the variegated feathers of a peacock, or various precious crystals, or the pointedness of a thorn. Some conceive being to be nirvana, some non-being, while others conceive that all things and nirvana are not to be distinguished from one another. Some, thinking that time is the creator and that as the rise of the world depends on time, they conceive that nirvana consists in the recognition of time as nirvana. Some think that there will be nirvana when the "twenty-five" truths are generally accepted, or when the king observes the six virtues, and some religionists think that nirvana is the attainment of paradise.

These views severally advanced by the philosophers with their various reasonings are not in accord with logic nor are they acceptable to the wise. They all conceive nirvana dualistically and in some causal connection; by these discriminations philosophers imagine nirvana, but where there is no rising and no disappearing, how can there be discrimination? Each philosopher, relying on his own textbook from which he draws his understanding, sins against the truth, because truth is not where he imagines it to be. The only result is that it sets his mind to wandering about and becoming more confused as nirvana is not to be found by mental searching, and the more his mind becomes confused the more he confuses other people.

As to the notion of nirvana as held by disciples and masters who still cling to the notion of an ego-self, and who try to find it by going off by themselves into solitude: their notion of nirvana is an eternity of bliss like the bliss of the Samadhis—for themselves. They recognize that the world is only a manifestation of mind and that all discriminations are of the mind, and so they forsake social relations and practice various spiritual disciplines and in solitude seek self-realization of Noble Wisdom by self-effort. They follow the stages to the sixth and attain the bliss of the Samadhis, but as they are still clinging to egoism they do not attain the "turning-about" at the deepest seat of consciousness and, therefore, they are not free from the thinking-mind and the accumulation of its habit-energy. Clinging to the bliss of the Samadhis, they pass to their nirvana, but it is not the nirvana of the Tathagatas. They are of those who have "entered the stream"; they must return to this world of life and death.

Then said Mahamati to the Blessed One: When the Bodhisattvas yield up their stock of merit for the emancipation of all beings, they become spiritually one with all animate life; they themselves may be purified, but in others there yet remain unexhausted evil and unmatured karma. Pray tell us, Blessed One, how the Bodhisattvas are given assurance of nirvana? and what is the nirvana of the Bodhisattvas?

The Blessed One replied: Mahamati, this assurance is not an assurance of numbers nor logic; it is not the mind that is to be assured but the heart. The Bodhisattva's assurance comes with the unfolding insight that follows passion hindrances cleared away, knowledge hindrance purified, and egolessness clearly perceived and patiently accepted. As the mortal-mind ceases to discriminate, there is no more thirst for life, no more sex-lust, no more thirst for learning, no more thirst for eternal life, with the disappearance of these fourfold thirsts, there is no more accumulation of habit-energy; with no more accumulation of habit-energy the defilements on the face of Universal Mind clear away, and the Bodhisattva attains self-realization of Noble Wisdom that is the heart's assurance of nirvana.

There are Bodhisattvas here and in other Buddha-lands, who are sincerely devoted to the Bodhisattva's mission and yet who cannot wholly forget the bliss of the Samadhis and the peace of nirvana—for themselves. The teaching of nirvana in which there is no substrate left behind, is revealed according to a hidden meaning for the sake of these disciples who still cling to thoughts of nirvana for themselves, that they may be inspired to exert themselves in the Bodhisattva's mission of emancipation for all beings. The Transformation-Buddhas teach a doctrine of nirvana to meet conditions as they find them, and to give encouragement to the timid and selfish. In order to turn their thoughts away from themselves and to encourage them to a deeper compassion and more earnest zeal for others, they are given assurance as to the future by the sustaining power of the Buddhas of Transformation, but not by the Dharmata-Buddha.

The Dharma which establishes the Truth of Noble Wisdom belongs to the realm of the Dharmata-Buddha. To the Bodhisattvas of the seventh and eighth stages, Transcendental Intelligence is revealed by the Dharmata-Buddha and the Path is pointed out to them which they are to follow. In the perfect self-realization of Noble Wisdom that follows the inconceivable transformation death of the Bodhisattva's individualized will-control, he no longer lives unto himself, but the life that he lives thereafter is the Tathagata's universalized life as manifested in its transformations. In this perfect self-realization of Noble Wisdom the Bodhisattva realizes that for Buddhas there is no nirvana.

The death of a Buddha, the great Parinirvana, is neither destruction nor death, else would it be birth and continuation. If it were destruction, it would be an effect-producing deed, which it is not. Neither is it a vanishing nor an abandonment, neither is it attainment, nor is it of no attainment; neither is it of one significance nor of no significance, for there is no nirvana for the Buddhas.

The Tathagata's nirvana is where it is recognized that there is nothing but what is seen of the mind itself; is where, recognizing the nature of the self-mind, one no longer cherishes the dualisms of discrimination; is where there is no more thirst nor grasping; is where there is no more attachment to external things. Nirvana is where the thinking-mind with all its discriminations, attachments, aversions, and egoism is forever put away; is where logical measures, as they are seen to be inert, are no longer seized upon; is where even the notion of truth is treated with indifference because of its causing bewilderment; is where, getting rid of the four propositions, there is insight into the abode of Reality. Nirvana is where the twofold passions have subsided and the twofold hindrances are cleared away and the twofold egolessness is patiently accepted; is where, by the attainment of the "turning-about" in the deepest seat of consciousness, self-realization of Noble Wisdom is fully entered into—that is the nirvana of the Tathagatas.

Nirvana is where the Bodhisattva stages are passed one after another; is where the sustaining power of the Buddhas upholds the Bodhisattvas in the bliss of the Samadhis; is where compassion for others transcends all thoughts of self; is where the Tathagata stage is finally realized.

Nirvana is the realm of Dharmata-Buddha; it is where the manifestation of Noble Wisdom that is Buddhahood expresses itself in Perfect Love for all; it is where the manifestation of Perfect Love that is Tathagatahood expresses itself in Noble Wisdom for the enlightenment of all—there, indeed, is nirvana!

There are two classes of those who may not enter the nirvana of the Tathagatas: there are those who have abandoned the Bodhisattva ideals, saying, they are not in conformity with the sutras, the codes of morality, nor with emancipation. Then there are the true Bodhisattvas who, on account of their original vows made for the sake of all beings, saying, "So long as they do not attain nirvana, I will not attain it myself," voluntarily keep themselves out of nirvana. But no beings are left outside by the will of the Tathagatas; some day each and every one will be influenced by the wisdom and love of the Tathagatas of Transformation to lay up a stock of merit and ascend the stages. But, if they only realized it, they are already in the Tathagata's nirvana for, in Noble Wisdom, all things are in nirvana from the beginning.

The following is a translation from a Sanskrit document whose author is unknown. He was probably a man of the first century of the Christian era. In the eighth century the work was carried from southern India to China, where it became a universally revered document in Chinese Buddhism. The first step, we learn, toward an understanding of the eternal verities must come through generosity, right living, and right thinking. As in all the central documents of Buddhism, purity and devotion must come first—before the higher levels of cosmic awareness can be achieved.

## The Surangama Sutra*

Then Ananda, rising in the midst of the assembly, straightened his robe, with the palms of his hands pressed together, knelt before the Lord Buddha. In the depths of his nature he was already enlightened and his heart was filled with happiness and compassion for all sentient beings and, especially, did he desire to benefit them by his newly acquired wisdom. He addressed the Lord Buddha, saying: Oh my Lord of Great Mercy! I have now seen the True Door for the attainment of Enlightenment, and have no more doubt about its being the only Door to Perfect Enlightenment. My Lord has taught us that those who are only starting the practice of Buddhahood and have not yet delivered themselves, but who already wish to deliver others, that this is a sign of Buddhahood. . . . Although I have not yet delivered myself, I already wish to deliver all sentient beings of this present existence. . . . What can I do to help them arrange a True Altar to Enlightenment within their minds so that they may be kept far away from all deceiving temptations and in whose progress there shall be no retrogression or discouragement in the attainment of Enlightenment?

In response to this appeal, the Blessed One addressed the assembly: Ananda has just requested me to teach how to arrange a True Altar of Enlightenment to which sentient beings may come for deliverance and protection. Listen carefully as I explain it to you.

Ananda and all in this assembly! In explaining to you the rules of the Discipline, I have frequently emphasized three good lessons, namely, (1) the only way to keep the Precepts is first to be able to concentrate the mind; (2) by keeping the Precepts you will be able to attain Samadhi [mystical ecstasy]; (3) by means of Samadhi one develops intelligence and wisdom. Having learned these three good lessons, one has gained freedom from the intoxicants and hindrances.

Ananda, why is concentration of mind necessary before one can keep the Precepts? And why is it necessary to keep the Precepts before

* From Lewis Browne (ed.), *The World's Great Scriptures* (New York: The Macmillan Company, 1961). Copyright 1946 by Lewis Browne. Reprinted by permission of The Macmillan Company.

one can attain Samadhi? And why is the attainment of Samadhi necessary before one may attain true intelligence and wisdom? Let me explain this to you. All sentient beings in all the six realms of existence are susceptible to temptations and allurements. As they yield to these temptations and allurements, they fall into and become fast bound to the recurring cycles of deaths and rebirths. Being prone to yield to these temptations and allurements, one must, in order to free himself from their bondage and their intoxication, concentrate his whole mind in a resolution to resist them to the uttermost. The most important of these allurements are the temptations to yield to sexual thoughts, desires, and indulgence, with all their following waste and bondage and suffering. Unless one can free himself from this bondage and these contaminations and exterminate these sexual lusts, there will be no escape from the following suffering, nor hope of advancement to enlightenment and peacefulness. No matter how keen you may be mentally, no matter how much you may be able to practice concentration, no matter to how high a degree of apparent Samadhi you may attain, unless you have wholly annihilated all sexual lusts, you will ultimately fall into the lower realms of existence. . . .

Therefore, Ananda, a man who tries to practice concentration without first attaining control of his lusts is like a man trying to bake bread out of a dough made of sand; bake it as long as he will, it will only be sand made a little hot. It is the same with sentient beings, Ananda. They cannot hope to attain Buddhahood by means of an indecent body. How can they hope to attain the wonderful experience of Samadhi out of bawdiness? If the source is indecent, the outcome will be indecent; there will ever be a return to the never-ending recurrence of deaths and rebirths. Sexual lust leads to multiplicity; control of mind and Samadhi leads to enlightenment and the unitive life of Buddhahood. Multiplicity leads to strife and suffering; control of mind and concentration leads to the blissful peace of Samadhi and Buddhahood.

Inhibition of sexual thoughts and annihilation of sexual lusts is the path to Samadhi, and even the conception of inhibiting and annihilating must be discarded and forgotten. When the mind is under perfect control and all indecent thoughts excluded, then there may be a reasonable expectation for the Enlightenment of the Buddhas. Any other teaching than this is but the teaching of the evil Maras. This is my first admonition as to keeping the Precepts.

The next important hindrance and allurement is the tendency of all sentient beings of all the six realms of existence to gratify their pride of egoism. To gain this one is prone to be unkind, to be unjust and cruel, to other sentient beings. This tendency lures them into the bondage of deaths and rebirths, but if this tendency can be controlled they will no longer be lured into this bondage for right control of mind will enable them to keep the Precept of kindness to all animate life. The reason for practicing concentration and seeking to attain Samadhi is to escape from the suffering of life, but in seeking to escape from suffering ourselves, why should we inflict it upon others? Unless you can so control your minds that even the thought of brutal unkindness and killing is abhorrent, you will never be able to escape from the bondage of the world's life. No matter how keen you may be mentally, no matter how much you may be able to practice concentration, no matter to how high a degree of Samadhi you may attain, unless you have wholly annihilated all tendency to unkindness toward others, you will ultimately fall into the realms of existence where the evil ghosts dwell. . . .

You of this great Assembly ought to appreciate that those human beings who . . . kill sentient beings and eat the flesh . . . are not true disciples of Buddha. Therefore, Ananda, next to teaching the people of the last age to put away all sexual lust, you must teach them to put an end to all killing and brutal cruelty.

If one is trying to practice concentration and is still eating meat, he would be like a man closing his ears and shouting loudly and then asserting that he heard nothing. The more one conceals things, the more apparent they become. Pure and earnest monks and Saints, when walking a narrow path, will never so much as tread on the growing grass beside the path. How can a monk, who hopes to become a deliverer of others, himself be living on the flesh of other sentient beings?

Pure and earnest monks, if they are true and sincere, will never wear clothing made of silk, nor wear boots made of leather because it involves the taking of life. Neither will they indulge in eating milk or cheese because thereby they are depriving the young animals of that which rightly belongs to them. . . . To wear anything, or partake of anything for self-comfort, deceiving one's self as to the suffering it causes others or other sentient life, is to set up an affinity with that lower life which will draw them toward it. So all monks must be very careful to live in all sincerity, refraining from even the appearance of unkindness to other life. Even in one's speech and especially in one's teaching, one must practice kindness, for no teaching that is unkind can be the true teaching of Buddha. Unkindness is the murderer of the life of Wisdom. This is the second admonition of the Lord Buddha as to the keeping of the Precepts.

Then there is the Precept of not taking anything that does not rightfully belong to one, not coveting it or even admiring it. One must learn to keep this Precept in all sincerity if he is to hope for escape from the chain of deaths and rebirths. The purpose of your practice of concentration is to escape from the suffering of this mortal life. No matter how keen you may be mentally, no matter how much you may be able to practice concentration, no matter to how high a degree of apparent ecstasy you may attain, unless you refrain from covetousness and stealing, you will fall into the realm of heretics. . . .

For all these various reasons, I teach my *bhikshu*-brothers not to covet comforts and privileges, but to beg their food, not here and there, or now and then, but to make it a regular habit so that they will be better able to overcome the greediness and covetousness that hinder their progress toward enlightenment. I teach them not to cook their own food even, but to be dependent upon others for even the poorest living so that they will realize their oneness with all sentient life and are but sojourners in this triple world. . . .

If any of my disciples who are trying to practice concentration do not abstain from stealing and covetousness, their efforts will be like trying to fill a leaking pot with water; no matter how long they try, they will never succeed. So all of you, my *bhikshu* disciples, with the exception of your poor garments and your begging bowls should have nothing more in possession. Even the food that is left over from your begging after you have eaten should be given to hungry sentient beings and should not be kept for the next meal. Moreover, you should look upon your own body, its flesh, blood, and bone, as not being your own but as being one with the bodies of all other sentient beings and so be ever ready to sacrifice it for the common need. Even when men beat you and scold you, you must accept it patiently and, with hands pressed together, bow to them humbly. Furthermore, you should not accept one teaching, or one principle, that is easy and agreeable, and reject the rest of the Doctrine; you should accept all with equitable mind lest you misinterpret the Doctrine to the new converts. Thus living, the Lord Buddha will confirm your attainment as one who has acquired the true Samadhi. As you teach the Doctrine to others, be sure that your teaching is in agreement with the above so that it may be regarded as a true teaching of Buddha, otherwise it would be as heretical as the deceptive words of the goblin-heretics who are murderers of the life of Wisdom. This is the third admonition of the Lord Buddha as it relates to the Precepts.

Then there is the Precept of not deceiving nor telling lies. If the sentient beings of the six realms of existence should refrain from killing, stealing, and adultery, and should refrain from even thinking about them, but should fail to keep the Precept of truthfulness . . . there would be no emancipation for them; they would . . . become prejudiced and egoistically assertive, and . . . lose their seed of Buddhahood. . . . They not only lose their own seed of Buddhahood, they destroy the seed of Buddhahood in others. Such disciples progressively lose their nature of kindness and gradually lose the measure of understanding that they had attained and shall at last sink into the Sea of the Three Kinds of Suffering, namely, (1) the suffering of pain, (2) the loss of enjoyment, (3) the suffering of decay. They will not attain to Samadhi for a long, long time in after lives.

I urge all Saints and holy men to choose to be reborn in order to deliver all sentient beings. You should make use of all manner of transformations, such as disciples, laymen, kings, lords, ministers, virgins, boy-eunuchs, and even as harlots, widows, adulterers, thieves, butchers, pedlers, etc., so as to be able to mingle with all kinds of people and to make known the true emancipation of Buddhism and the following peace of Samadhi. . . . To teach the world to observe the Precept of truthful sincerity, to practice concentration with sincerity and to attain a true ecstasy, this is the clear and true instruction of the Lord Buddha.

Therefore, Ananda, if any disciple does not abstain from deceit, he is like a man molding human dung instead of carving sweet-smelling sandalwood. . . . But disciples whose lives are as straight as the chord of a bow will certainly attain Samadhi. They need never fear the wiles of the Maras. They are the disciples who are certain to attain the Savior's supreme understanding and insight. Any lesson or instruction that is in agreement with the foregoing can be relied upon as being a true teaching of the Lord Buddha. Differing from it, it is simply a false teaching of the heretics who have always been murderers of the Life of Wisdom. This is the fourth admonition of the Lord Buddha.

Ananda! As you have asked me as to the best method for concentrating the mind of those who have difficulty in following the common methods, I will now reveal to you the Lord Buddha's Secret Method for the attainment of saviorhood. But you must remember that it is of first importance to fully observe the Four Precepts as explained above.

# Two

# THE PSYCHOLOGY OF
# CHINA

## INTRODUCTION

From the Western viewpoint India and China have much in common. Their huge land masses permit easy movement and communication within the boundaries of these countries. They possess large regions of arable land, and the capacity to feed very large numbers of human beings. They have maintained a high degree of physical and cultural isolation through most of their history—both from each other and from the rest of the world.

But their psychologies are distinctive indeed. It would not help us much to try to make them coalesce by speaking of them as examples of a general "Asian psychology" or to speak in broad terms about the "psychology of the East." We have already seen that Indian psychology went through extraordinary developments from about 1500 B.C. to about A.D. 500, and that even at any one point in time India possessed many sharply contrasting psychological systems. To begin our brief consideration of the psychology of China, it will be best not to look constantly for parallels with India, but rather to start completely afresh, and bring in the parallels and the known instances of intercommunication between Indian and Chinese thinkers only after we have looked at the idiom and basic pattern of the Chinese approach.

A rich and complex civilization already existed in China over a thousand years before the birth of Christ. Archaeological evidence and early historical records reveal art works of great power and beauty, a great social organization involving a well-developed family system, and a sustained movement toward political unification. It is to the first millennium before Christ that we must look for the first historical picture of a successful agricultural system, a system of government operated by highly selected and trained political officials, and a class of teachers held in high repute who served the state by recruiting and training each new generation. It is in this setting of officialdom, the world of royal court and provincial deputy administrators, that the teaching system achieved its glory as a device for conserving classical literary products and at the same time encouraging philosophical speculation about the nature of the world and of man, and, above all, established the system of moral ideas which led to the stabilization and improvement of the official duties of the functionary class in which this type of thinking prevailed.

As will be pointed out more fully by Professor Francis Hsu (pages 145 ff.), the primary burden of the philosophy that developed among the teaching class, the philosopher-officials, was directed toward practical, effective administration. The closest parallel with India is in the court documents we have presented above (page 72). In China it was not a secondary task of philosophy to solve everyday problems; it was rather the primary task. We shall see some exceptions, but we must keep them in just perspective. The earliest Chinese philosophy—and the earliest Chinese psychology embedded within it—is a psychology of everyday human nature: how it apprehends and uses the physical world and especially the human world around it, how it guides the way to wisdom and toward decisions which need not be regretted or revoked.

If the reader finds in this chapter a lack of that speculative philosophy pursued for its own delight in *ultimate* realities which we encounter among the Greeks and among the Indians, it is not because we have not looked for it. One may, as a matter of fact, find a considerable amount of speculative philosophy in China, but with this marked reservation: speculative philosophies imported from India were assimilated, repeated, taught, and evidently enjoyed, but they were hardly developed at all. When Buddhism, for example, was imported into China, it early achieved its characteristic Chinese expression, which then remained stabilized and essentially unchanging century after century to a degree which permits us to challenge all assertions regarding any indigenous

Chinese speculative psychological system. The Chinese have a great deal to offer, but their contribution lies in a different direction.

## THE BOOK OF CHANGES

We begin our sketch of Chinese psychology with the I Ching, the Book of Changes, a huge compendium of observations on the world and on human nature, which reads much like the proverbs or wise sayings with which we in the West are familiar. The proverbs, however, must be made to serve man in an immediate, practical way, just as the Delphic oracle served the Greeks when confronted with crises and just as the Romans relied on auspices, the observation of the entrails of birds, as clues to what might happen, and as indications of the right course to follow in each contingency. How can abstract proverbs such as "One man's meat is another man's poison" or "Sauce for the goose is sauce for the gander" help one with a problem in love, in health, in material fortune, or in political wisdom? One can make them serve these purposes, thought the Chinese, by assuming that there is a natural order in the world which will reveal itself if one gives up trying to control it. In other words, one utilizes a method of "chance," as we do when we throw the dice or toss a coin. Some inscrutable principle in the cosmos will make the die come up three or five, and this, according to the common man's logic, will allow some principle of abstract order, justice, or good sense to replace the capricious or arbitrary decisions of men. At times there is even an explicit supernatural element forced into service, as when one opens the Bible at random and, with eyes averted, puts his thumb on a particular verse, where he may read: "Arise and go toward the south." If he had been trying to decide whether the family finances justified a trip to Florida, the solution is there. Among many of the ancient Chinese who used the Book of Changes there may have been a feeling that if one randomly selects a specific proverb, the event is something of a symbol of what the universe as a whole portends to him. The little is symbolic of the big. Of all possible wise utterances that might apply to human affairs, the one that is hit upon at this moment is symbolic of what is cosmically appropriate.

The universe is conceived by the I Ching to consist of various ways of ordering the active, continuous, or male (———) principle and the passive, discontinuous, or female (— —) principle; and the complex patterns in which these may be arranged, such as

$$\begin{array}{ccc} \overline{\phantom{xx}} \\ \overline{\phantom{xx}} & \text{or} & \overline{\phantom{x}}\ \overline{\phantom{x}} \\ \overline{\phantom{x}}\ \overline{\phantom{x}} & & \overline{\phantom{xx}} \end{array}$$

are symbolic of the realities, especially the processes of change, which make up the universe. These are the trigrams. The corresponding patterns of six groups of lines are hexagrams:

We find William D. Kennedy's explanation useful:

> You address a question to the I Ching as if it were a person and toss a bunch of yarrow sticks or three coins. The first three throws establish the lower trigram and the next three the upper. The combination determines the hexagram and the book is turned to the page where this hexagram is to be found. It might bear such a title as "The Marrying Maiden" or "Possession in Great Measure." The interpretation is given in poetic, symbolic, and philosophical terms under the headings of The Judgement, The Image, and The Lines, the latter consisting of special interpretations of key throws, some of which may send the thrower to another hexagram (The Change) for further elucidation.*

The I Ching is thus a book of oracles. The oracular statements are, however, based on general principles of the universe that

* William D. Kennedy, *Psychology: The Reluctant Science,* draft of August 1, 1963 (private distribution only), p. 15.

affect human life. Some sensible and useful ideas about process, types of change, and so forth are stated. One might say that the book is a fascinating combination of sense and nonsense.

More or less as the random formation of tea leaves allows one to read into them many interpretations, so the random arrangement of yarrow sticks when the bundle is split is interpreted according to a set of principles which in one way sound like the Delphic oracle at its shrewdest and in another way sound like a fortuneteller's emptiest babblings. But throughout the long history of China a rich practical use of the I Ching has been made, and the assumptions contained in the I Ching and the commentary upon it have abundant wisdom.

We use here the translation and commentary by Richard Wilhelm.* Wilhelm notes in his introduction to his edited version of the book that both Confucianism (pages 146 ff.) and Taoism (pages 155 ff.) have their roots in the I Ching, and that it alone, among all the Confucian classics, escaped the great burning of the books under Ch'in Shih Huang Ti. "Even the policy makers of so modern a state as Japan, distinguished for their astuteness, do not scorn to refer to it for counsel in difficult situations. . . . Apart from [its] mechanistic number mysticism, a living stream of deep human wisdom was constantly flowing through the channel of this book into everyday life, giving to China's great civilization that ripeness of wisdom, distilled through the ages, which we wistfully admire in the remnants of this last truly autochthonous culture."

Wilhelm's introduction includes an explanation of the arithmetical patterns of the hexagrams. The answer "yes" was indicated by a simple unbroken line: ———. Further differentiation was made possible by using two shorter lines: —— ——. Later came eight "trigrams," shown on page 138, which "were conceived as images of all that happens in heaven and on earth. At the same time, they were held to be in a state of continual transition, one changing into another, just as transition from one phenomenon to

---

* All quotations are from *The I Ching, or Book of Changes*, the Richard Wilhelm translation from Chinese into German, rendered into English by Cary F. Baynes, with a Foreword by C. G. Jung (2nd ed.; New York: Bollingen Foundation, 1966). Second edition copyright 1966 by Bollingen Foundation; distributed by Pantheon Books.

another is continually taking place in the physical world. Here we have the fundamental concept of the Book of Changes. The eight trigrams are symbols standing for changing transitional states; they are images that are constantly undergoing change. Attention centers not on things in their state of being—as is chiefly the case in the Occident—but upon their movements in change. The eight trigrams therefore are not representations of things as such but of their tendencies in movement."

Wilhelm also notes that "in addition to the law of change and to the images of the states of change . . . another factor to be considered is the course of action. Each situation demands the action proper to it. In every situation, there is a right and a wrong course of action. Obviously, the right course brings good fortune and the wrong course brings misfortune. Which, then, is the right course in any given case? This question was the decisive factor." The fact that this seems so obvious to us as hardly to be worth saying obscures the implication that awareness of the consequences of one's own action came slowly in the intellectual development of man.

Wilhelm explains that in the course of the evolution of the I Ching, King Wen, about 1150 B.C. brought about a change. King Wen and his son "endowed the hitherto mute hexagrams and lines . . . with definite counsels for correct conduct." Thus the individual came to share in shaping fate. His actions intervened as determining factors in world events, the more decisively so, that earlier he was able with the aid of the Book of Changes to recognize situations in their germinal phases.

"The only thing about all this that seems strange to our modern sense is the method of learning the nature of a situation through the manipulation of yarrow stalks. This procedure was regarded as mysterious, however, simply in the sense that the manipulation of the yarrow stalks makes it possible for the unconscious in man to become active."

But the oracular and, to us, sheer magical use of the book is, Wilhelm felt, only part of its potentiality: "Of far greater significance than the use of the Book of Changes as an oracle is its other use, namely, as a book of wisdom. Lao-tzu knew this book, and some of his profoundest aphorisms were inspired by it. . . . Confucius too knew the Book of Changes and devoted himself to reflection upon it. . . . The Book of Changes as edited and annotated by Confucius is the version that has come down to our time."

A central concern of wisdom is the role of change: "The underlying idea of the whole is the idea of change. It is related in the Analects that Confucius, standing by a river, said: 'Everything flows on and on like this river, without pause, day and night.' This expresses the idea of change. He who has perceived the meaning of change fixes his attention no longer on transitory individual things but on the immutable, eternal law at work in all change. This law is the Tao of Lao-tsu, the course of things, the principle of the one in the many. . . . This fundamental postulate is the 'great primal beginning' of all that exists. . . . Later Chinese philosophers devoted much thought to this idea of a primal beginning. A still earlier beginning, *wu chi*, was represented by the symbol of the circle. Under this conception, *t'ai chi* was represented by the circle divided into the light and the dark, yang and yin." In this a world of opposites is posited. "These opposites became known under the names yin and yang and created a great stir . . . in the centuries just before our era. . . . In its primary meaning yin is 'the cloudy,' 'the overcast,' and yang means actually . . . something 'shone upon,' or bright. . . . The world of being arises out of their change and interplay. . . . Thus change is conceived of partly as the continuous transformation of the one force into the other and partly as a cycle of complexes of phenomena, in themselves connected, such as day and night, summer and winter."

Here the principles of multiplicity of factors, of contrasts of opposites, and the changes resulting from the interplay of them foreshadow such concepts as multiple determinism, dualism, and conflict in psychoanalytic thought.

"The eight trigrams are images not so much of objects as of states of change. This view is associated with the concept expressed in the teachings of Lao-tzu, as also in those of Confucius, that every event in the visible world is the effect of an 'image,' that is, of an idea in the unseen world. . . . This theory of ideas is applied in a twofold sense. The Book of Changes shows the images of events and also the unfolding of conditions *in statu nascendi*. . . . The third element fundamental to the Book of Changes is the judgments. The judgments clothe the images in words, as it were; they indicate whether a given action will bring good fortune or misfortune, remorse or humiliation. . . . In its judgments, and in the interpretations attached to it from the time of Confucius on, the Book of Changes opens to the reader the richest treasure of Chinese wisdom; at the same time it affords him a comprehensive view of the varieties of human experience, enabling him thereby to shape his life of his own sovereign will into an organic whole and so to direct it that it comes into accord with the ultimate Tao lying at the root of all that exists."

There are, for example, hexagrams on the subject of "Youthful Folly," "The Army," "Peace," and "Standstill (Stagnation)."

The hexagram dealing with "Peace" contains a symbol for the receptive, which stands above, and the creative, which moves upward. Their influences meet and are in harmony, and as a result all things bloom and prosper. Wilhelm says: "This hexagram denotes a time in nature when heaven seems to be on earth. Heaven has placed itself beneath the earth, and so their powers unite in deep harmony. Then peace and blessing descend upon all living things.

"In the world of man it is a time of social harmony; those in high places show favor to the lowly, and the lowly and inferior in their turn are well disposed toward the highly placed. There is an end to all feuds."

> "No plain not followed by a slope.
> No going not followed by a return.
> He who remains persevering in danger
> Is without blame.
> Do not complain about this truth;
> Enjoy the good fortune you still possess."

Wilhelm's commentary on this passage reads: "Everything on earth is subject to change. Prosperity is followed by decline: this is the eternal law on earth. Evil can indeed be held in check but not permanently abolished. It always returns. This conviction might induce melancholy, but it should not; it ought only to keep us from falling into illusion when good fortune comes to us. If we continue mindful of the danger, we remain persevering and make no mistakes. As long as a man's inner nature remains stronger and richer than anything offered by external fortune, as long as he remains inwardly superior to fate, fortune will not desert him."

The hexagram relating to "Standstill (Stagnation)" is just the opposite of the preceding one. Heaven and earth are in bad relationships. Heaven draws further away and earth sinks further into the depths. This hexagram is linked with the fall of the year, when the zenith has been passed and the decay of autumn is setting in.

As in the case of other great texts, there were commentators who undertook the task of explaining and elaborating the basic ideas in the I Ching. The Ta Chuan is such a commentary: "The Great Treatise." This is divided into parts and chapters. The first part deals with underlying principles; the first chapter deals with the changes in the Universe and in the Book of Changes. Here we can quote the text (printed in italic type) and Wilhelm's commentary on the first eight selections from the text.

## The Changes in the Universe and in the Book of Changes

1. *Heaven is high, the earth is low; thus the Creative and the Receptive are determined. In correspondence with this difference between low and high, inferior and superior places are established.*

   *Movement and rest have their definite laws; according to these, firm and yielding lines are differentiated.*

   *Events follow definite trends, each according to its nature. Things are distinguished from one another in definite classes. In this way good fortune and misfortune come about. In the heavens phenomena take form; on earth shapes take form. In this way change and transformation become manifest.*

In the Book of Changes a distinction is made between three kinds of change: non-change, cyclic change, and sequent change. Non-change is the background, as it were, against which change is made possible. For in regard to any change there must be some fixed point to which the change can be referred; otherwise there can be no definite order and everything is dissolved in chaotic movement. This point of reference must be established, and this always requires a choice and a decision. It makes possible a system of coordinates into which everything else can be fitted. Consequently, at the beginning of the world, as at the beginning of thought, there is the decision, the fixing of the point of reference. Theoretically any point of reference is possible, but experience teaches that at the dawn of consciousness one stands already inclosed within definite, prepotent systems of relationships. The problem then is to choose one's point of reference so that it coincides with the point of reference for cosmic events. For only then can the world created by one's decision escape being dashed to pieces against prepotent systems of relationships with which it would otherwise come into conflict. Obviously the premise for such a decision is the belief that in the last analysis the world is a system of homogeneous relationships—that it is a cosmos, not a chaos. This belief is the foundation of Chinese philosophy, as of all philosophy. The ultimate frame of reference for all that changes is the non-changing.

The Book of Changes takes as the foundation for this system of relationships the distinction between heaven and earth. There is heaven, the upper world of light, which, though incorporeal, firmly regulates and determines everything that happens, and over against heaven there is the earth, the lower, dark world, corporeal, and dependent in its movements upon the phenomena of heaven. With this differentiation of above and below there is posited, in one way or another, a difference in value, so that the one principle, heaven, is the more exalted and honored, while the other, earth, is regarded as lesser and lower. These two cardinal principles of all existence are then symbolized in the two fundamental hexagrams of the Book of Changes, *the creative* and *the receptive*. In the last analysis, this cannot be called a dualism. The two principles are united by a relation based on homogeneity; they do not combat but complement each other. The difference in level creates a potential, as it were, by virtue of which movement and living expression of energy become possible.

This association of high and low with value differentiations leads to the differentiation of superior and inferior. This is expressed symbolically in the hexagrams of the Book of Changes, which are considered to have high and low, superior and inferior places. Each hexagram consists of six places, of which the odd-numbered ones are superior and the even-numbered ones inferior.

There is another difference bound up with this one. In the heavens constant movement and change prevail; on earth fixed and apparently lasting conditions are to be observed. On closer scrutiny, this is only delusion. In the philosophy of the Book of Changes nothing is regarded as being absolutely at rest; rest is merely an intermediate state of movement, or latent movement. However, there are points at which the movement becomes visible. This is symbolized by the fact that the hexagrams are built up of both firm and yielding lines. The firm, the strong, is designated as the principle of movements, the yielding as the principle of rest. The firm is represented by an undivided line, corresponding with the light principle, the yielding by a divided line that corresponds with the dark principle.

The fact that the character of the line (firm, yielding) combines with the character of the place (superior, inferior) results in a great multiplicity of possible situations. This serves to symbolize a third nexus of events in the world. There are conditions of equilibrium, in which a certain harmony prevails, and conditions of disturbed equilibrium, in which confusion prevails. The reason is that there is a system of order pervading the entire world. When, in accordance with this order, each thing is in its appropriate place, harmony is established. Such a tendency toward order can be observed in nature. The places attract related elements, as it were, so that harmony may come about. However, a parallel tendency is also at work. Not only are things determined by their tendency toward order: they move also by virtue of forces imparted to them, so to speak, mechanically from the outside. Hence it is not possible for equilibrium to be attained under all circumstances, for deviations may occur, bringing with them confusion and disorder. In the sphere of human affairs, the condition of harmony assures good fortune, that of disharmony predicates misfortune. These complexes of occurrences can be represented by the combinations of lines and places, as pointed out above.

Another law is to be noted. Owing to changes of the sun, moon, and stars, phenomena take form in the heavens. These phenomena obey definite laws. Bound up with them, shapes come into being on earth, in accordance with identical laws. Therefore the processes on earth—blossom and fruit, growth and decay—can be calculated if we know the laws of time. If we know the laws of change, we can precalculate in regard to it, and freedom of action thereupon becomes possible. Changes are the imperceptible tendencies to divergence that, when they have reached a certain point, become visible and bring about transformations.

These are the immutable laws under which, according to Chinese thought, changes are consummated. It is the purpose of the Book of Changes to demonstrate these laws by means of the laws of change operating in the respective hexagrams. Once we succeed in completely reproducing these laws, we acquire a comprehensive view of events; we can understand past and future equally well and bring this knowledge to bear in our actions.

*2. Therefore the eight trigrams succeed one another by turns, as the firm and the yielding displace each other.*

Here cyclic change is explained. It is a rotation of phenomena, each succeeding the other until the starting point is reached again. Examples are furnished by the course of the day and year, and by the phenomena that occur in the organic world during these cycles. Cyclic change, then, is recurrent change in the organic world, whereas sequent change means the progressive [non-recurrent] change of phenomena produced by causality.

The firm and the yielding displace each other within the eight trigrams. Thus the firm is transformed, melts as it were, and becomes the yielding; the yielding changes, coalesces, as it were, and becomes the firm. In this way the eight trigrams change from one into another in turn, and the regular alternation of phenomena within the year takes its course. But this is the case in all cycles, the life cycle included. What we know as day and night, summer and winter—this, in the life cycle, is life and death.

To make more intelligible the nature of cyclic change and the alternations of the trigrams produced by it, their sequence in the Primal Arrangement is shown below. There are two directions of movement, the one rightward, ascending, the other backward, descending. The former starts from the low point, K'un, the Receptive, earth; the latter starts from the high point, Ch'ien, the Creative, heaven.

*3. Things are aroused by thunder and lightning; they are fertilized by wind and rain. Sun and moon follow their courses and it is now hot, now cold.*

Here we have the sequence of the trigrams in the changing seasons of the year, and in such a way that each is the cause of the one next following. Deep in the womb of earth there stirs the creative force, Chên, the Arousing, symbolized by thunder. As this electrical force appears there are formed centers of activation that are then discharged in lightning. Lightning is Li, the Clinging, flame. Hence thunder is put before lightning. Thunder is, so to speak, the agent evoking the lightning; it is not merely the sounding thunder. Now the movement shifts; thunder's opposite, Sun, the wind, sets in. The wind brings rain, K'an. Then there is a new shift. The trigrams Li and K'an, formerly acting in their secondary forms as lightning and rain, now appear in their primary forms as sun and moon. In their cyclic movement they cause cold and heat. When the sun reaches the zenith, heat sets in, symbolized by the trigram of the southeast, Tui, the Joyous, the lake. When the moon is at its zenith in the sky, cold sets in, symbolized by the trigram of the northwest, Kên, the mountain, Keeping Still.

4. *The way of the Creative brings about the male. The way of the Receptive brings about the female.*

Here the beginning of sequent change appears, manifested in the succession of the generations, an onward-moving process that never returns to its starting point. This shows the extent to which the Book of Changes confines itself to life. For according to Western ideas, sequent change would be the realm in which causality operates mechanically; but the Book of Changes takes sequent change to be the succession of the generations, that is, still something organic.

The Creative, insofar as it enters as a principle into the phenomenon of life, is embodied in the male sex; the Receptive is embodied in the female sex. Thus the Creative in the lowest line of each of the sons (Chên, Li, Tui, in the Primal Arrangement), and the Receptive in the lowest line of each of the daughters (Sun, K'an, Kên, in the Primal Arrangement), is the sex determinant of the given trigram.

5. *The Creative knows the great beginnings.*
   *The Receptive completes the finished things.*

Here the principles of the Creative and the Receptive are traced further. The Creative produces the invisible seeds of all development. At first these seeds are purely abstract, therefore with respect to them there can be no action nor acting upon; here it is knowledge that acts creatively. While the Creative acts in the world of the invisible, with spirit and time for its field, the Receptive acts upon matter in space and brings material things to completion. Here the processes of generation and birth are traced back to their ultimate metaphysical meanings.

6. *The Creative knows through the easy.*
   *The Receptive can do things through the simple.*

The nature of the Creative is movement. Through movement it unites with ease what is divided. In this way the Creative remains effortless, because it guides infinitesimal movements when things are smallest. Since the direction of movements is determined in the germinal stage of being, everything else develops quite effortlessly of itself, according to the law of its nature.

The nature of the Receptive is repose. Through repose, the absolutely simple becomes possible in the spatial world. This simplicity, which arises out of pure receptivity, becomes the germ of all spatial diversity.

7. *What is easy, is easy to know; what is simple, is easy to follow. He who is easy to know attains fealty. He who is easy to follow attains works. He who possesses attachment can endure for long; he who possesses works can become great. To endure is the disposition of the sage; greatness is the field of action of the sage.*

This passage points out how the easy and the simple take effect in human life. What is easy is readily understood, and from this comes its power of suggestion. He whose ideas are clear and easily understood wins men's adherence because he embodies love. In this way he becomes free of confusing conflicts and disharmonies. Since the inner movement is in harmony with the environment, it can take effect undisturbed and have long duration. This consistency and duration characterize the disposition of the sage.

It is exactly the same in the realm of action. Whatever is simple can easily be imitated. Consequently, others are ready to exert their energy in the same direction; everyone does gladly what is easy for him, because it is simple. The result is that energy is accumulated, and the simple develops quite naturally into the manifold. Thus it grows, and the sage's mission to lead the multitude to the performance of great works is fulfilled.

8. *By means of the easy and the simple we grasp the laws of the whole world. When the laws of the whole world are grasped, therein lies perfection.*

Here we are shown how the fundamental principles demonstrated above are applied in the Book of Changes. The easy and the simple are symbolized by very slight changes in the individual lines. The divided lines become undivided lines as the result of an easy movement that joins their separated ends; undivided lines become divided ones by means of a simple division in the middle. Thus the laws of all processes of growth under heaven are depicted in these easy and simple changes, and thereby perfection is attained.

Hereby the nature of change is defined as change of the smallest parts. This is the fourth meaning of the Chinese word *I*— a connotation that has, it is true, only a loose connection with the meaning "change."

It is interesting to see that the essence of continuing life itself is considered to consist in these oppositions; if this active process that is life ceased, the oppositions would be obliterated and with them the processes of active change, and when all of this ceased the world would be dead. (Compare Heraclitus: "Strife is the father of all things.")

We are shown here how the individual can attain mastery over fate by means of the Book of Changes. Its principles contain the categories of all that is—literally, the molds and the scope of all transformations. These categories are in the mind of man; everything, all that happens and everything that undergoes transformation, must obey the laws prescribed by the mind of man. Not until these categories become operative do things become things. These categories are laid down in the Book of Changes; hence it enables us to penetrate and understand the movements of the light and the dark, of life and death, of gods and demons. This knowledge makes possible mastery over fate, because fate can be shaped if its laws are known. The reason why we can oppose fate is that reality is always conditioned, and these conditions of time and space limit and determine it. The spirit, however, is not bound by these determinants and can bring them about as its own purposes require. The Book of Changes is so widely applicable because it contains only these purely spiritual relationships, which are so abstract that they can find expression within every framework of reality. They contain only the Tao that underlies events. Therefore all chance contingencies can be shaped according to this Tao. The conscious application of these possibilities assures mastery over fate.

The second part of the commentary on I Ching deals with the way the principles outlined actually work. Chapter I has to do with the signs and lines, in relation to Creating and Acting.

Chapter II deals with the history of civilization. Development of evolution is implied in such statements as: "In the beginning there was as yet no moral or social order. Men knew their mothers only, not their fathers." Development in many aspects of human life and relationships is described, including the development of clans and of organization and government.

Chapter VII returns to processes of human functioning: the relation of certain hexagrams to character formation. There seems to be an implication of awareness of the relation between the condi-

THE PSYCHOLOGY OF CHINA

tion of the times and the concepts of the writer. Some of the relationships and their expressions seem more obscure than the ideas reviewed in the preceding sections. However, certain often repeated terms make very explicit what "character" involves. These include "good conduct," "modesty," which "honors others and thereby attains honor for itself"; "it regulates human intercourse in such a way that friendliness invokes friendliness." "Return" is regarded as contributing the capacity "to be able constantly to prevail in its own unique character against any temptation of the surroundings" and "also suggests lasting reform falling upon errors committed, and a self-examination and self-knowledge necessary for this." "Duration" brings about "firmness of character in the frame of time." From manifold movements and experiences, fixed rules are derived, so that a unified character results.

The concept of "decrease" refers to the decrease of the influence of the lower faculties, the untamed instinct, in favor of the higher life of the mind (the commentator notes that this is the essence of character training). In contemporary psychoanalytical terms we would see this as the equivalent of Freud's statement: "Where Id was, there shall Ego be." Two phrases are noted. First a difficult thing—the taming of the instinct—then the easy phase, when character is under control and harm is thus avoided.

The discussion of "increase" emphasizes the fact that fullness of character is valued; "mere ascetism is not enough to make a good character: greatness is also needed."

Chapter XII, which constitutes our next selection, is a summary of this commentary in which much wisdom is distilled into a few words. In the last item the wise comments are made that verbal expressions are related to inner states. A person who is conspiring revolt speaks in confused ways. Doubt tends to lead to ramifications of words. By contrast, the words of "men of good fortune" are concise. Excited people also talk too much, while slanderers of good people are indirect and roundabout or use innuendoes. The words of a person who has lost his stance are confused or "twisted." Clinical psychologists could hardly do better in interpreting the verbal style of a patient than to use some of the insights included here.

## Summary

1. *The Creative is the strongest of all things in the world. The expression of its nature is invariably the easy, in order thus to master the dangerous. The Receptive is the most devoted of all things in the world. The expression of its nature is invariably simple, in order thus to master the obstructive.*

The two cardinal principles of the Book of Changes, the Creative and the Receptive, are here once more presented in their essential features. The Creative is represented as strength, to which everything is easy, but which remains conscious of the danger involved in working from above downward, and thus masters the danger. The Receptive is represented as devotion, which therefore acts simply, but which is conscious of the obstructions inherent in working from below upward, and hence masters these obstructions.

2. *To be able to preserve joyousness of heart and yet to be concerned in thought: in this way we can determine good fortune and misfortune on earth, and bring to perfection everything on earth.*

Joyousness of heart is the way of the Creative. To be concerned in thought is the way of the Receptive. Through joyousness one gains an over-all view of good fortune and misfortune, through concern one attains the possibility of perfection.

3. *Therefore: The changes and transformations refer to action. Beneficent deeds have good auguries. Hence the images help us to know the things, and the oracle helps us to know the future.*

The changes refer to action. Hence the images of the Book of Changes are of such sort that one can act in accordance with the changes and know reality. Events tend toward good fortune or misfortune, which are expressed in omens. In that the Book of Changes interprets these omens, the future becomes clear.

4. *Heaven and earth determine the places. The holy sages fulfill the possibilities of the places. Through the thoughts of men and the thoughts of spirits, the people are enabled to participate in these possibilities.*

Heaven and earth determine the places and thereby the possibilities. The sages make these possibilities into reality, and through the collaboration of the thoughts of spirits and of men in the Book of Changes, it becomes possible to extend the blessings of culture to the people as well.

5. *The eight trigrams point the way by means of their images; the words accompanying the lines, and the decisions, speak according*

*to the circumstances. In that the firm and the yielding are inter-
spersed, good fortune and misfortune can be discerned.*

6. *Changes and movements are judged according to the furtherance
(that they bring). Good fortune and misfortune change according to
the conditions. Therefore: Love and hate combat each other, and
good fortune and misfortune result therefrom. The far and the near
injure each other, and remorse and humiliation result therefrom.
The true and the false influence each other, and advantage and
injury result therefrom. In all the situations of the Book of Changes
it is thus: When closely related things do not harmonize, misfortune
is the result: this gives rise to injury, remorse, and humiliation.*

The close relationships between the lines are those of correspondence
and of holding together. According to whether the lines attract or repel
one another, good fortune or misfortune ensues, in all the gradations
possible in each case.

There is an interesting foreword to the Wilhelm translation of
the I Ching by C. G. Jung, who took the book seriously and put
to it two questions to which he felt he received usable and intelligible answers. He regards the book as a "method of exploring the
unconscious." He says: "The moment under actual observation
appears to the ancient Chinese view more of a chance hit than a
clearly defined result of concurring causal chain processes. . . .
While the Western mind carefully sifts, weighs, selects, classifies,
isolates, the Chinese picture of the moment encompasses everything down to the minutest nonsensical detail, because all of the
ingredients make up the observed moment."

It is assumed that the fall of the coins or the result of the division of the bundle of yarrow sticks is what it necessarily must be
in a given "situation," inasmuch as anything happening in that
moment belongs to it as an indispensable part of the picture. If a
handful of matches is thrown to the floor, they form the pattern
characteristic of that moment. But such an obvious truth as this
reveals its meaningful nature only if it is possible to read the
pattern and to verify its interpretation partly by the observer's
knowledge of the subjective and objective situation, partly by the
character of subsequent events.

Jung's comment on the thought processes of the analyst indicates the basis for his sense of affinity to the mode of thinking in the I Ching: "Probably in no other field do we have to reckon with so many unknown quantities and nowhere else do we become more accustomed to adopting methods that work even though for a long time we may not know why they work. Unexpected cures may arise from questionable therapies and unexpected failures from allegedly reliable methods. In the exploration of the unconscious we come upon very strange things, from which a rationalist turns away with horror, claiming afterward that he did not see anything. The irrational fullness of life has taught me never to discard anything, even when it goes against all our theories (so short-lived at best) or otherwise admits of no immediate explanation. It is of course disquieting, and one is not certain whether the compass is pointing true or not; but security, certitude, and peace do not lead to discoveries. It is the same with this Chinese mode of divination. Clearly the method aims at self-knowledge, though at all times it has also been put to superstitious use."

❁

Based on a rich folklore of which the I Ching is an example, the formal religious philosophy of China may be considered to fall into three main divisions: Confucianism (sixth century B.C.), Taoism, the philosophy of Lao-tzu (same era), and Buddhism (imported from India in the early centuries of the Christian era).

The atmosphere most useful in approaching Chinese philosophy and psychology is suggested by our consultant, Professor Francis Hsu, of Northwestern University:

The Chinese do not explore the development or the processes of the mind very much. They have little concern with first and last things. Though ancient philosophers—Lao-tzu, Chuang-tzu, Confucius, Wang Yang-ming—were concerned with the relation between man, earth, heaven, and things, their main aim was usually to elucidate the "correct" relationship among men or to ascertain the best way of improving that relationship.

My original impression of the Tao Teh Ching—Lao-tzu's great poem of the sixth century B.C.—was that its primary interest was supernatural or mystical. This impression of mine was strengthened when I really read and understood the following passage in Chinese from the Tao Teh Ching (Section 25):

> It may be regarded as the Mother of the world.
> It does not know its name;
> I style it Tao.

During the years when I was raised and educated in China, I never had the encouragement or inducement to study the Tao Teh Ching at all. I simply entertained the prevailing popular Chinese notion that the Tao Teh Ching was mystical. I now see, however, that even this book is not supernaturally oriented in the main; that its major contents deal with social, political, and economic relationships; and that these thoughts had obviously had an effect on Confucius.

C. G. Jung, in his foreword to the I Ching, speaks as though Confucius and Lao-tzu had the same sort of view of man and the universe. This is a perceptive insight, but when Jung later complains that many extraneous matters, some of foreign origin, were associated with the I Ching and "forced Chinese philosophical thinking more and more into a rigid formalization," he is clearly wrong. He has failed to see the implications of the very Chinese approach he speaks of. In my view, the type of intellectual exercise exemplified by the I Ching is in accord with the primary Chinese orientation of life and invites the extraneous matters that Jung thinks have intruded into the Chinese system. In the Chinese approach to life, according to Jung, natural science could not flower. "Increasingly hair-splitting cabalistic speculations" were bound to "envelop the Book of Changes in a cloud of mystery" and "everything of the past and of the future" would be forced "into this system of numbers."

The Chinese orientation is *situation-centered*, which is rooted, from the family onward, in human relationships. That situation-centered orientation fits well with the philosophy of the I Ching. The I Ching's main purpose is to enable man to detect the effect or significance of a given situation or a given configuration to man: that situation or configuration includes all elements of the universe in addition to human beings.

## CONFUCIANISM

A complex and stable social order, expressing much political sagacity on the one hand and much artistic power and skill on the other, had been achieved well before 1000 B.C. This social order was not conceived to be the realization of the ideals of a great prophet or religious leader, but to be a natural expression of cosmic order and of human intelligence. There was plainly a place for a great prophetic figure who would accept and systematize the inherent ethics, law, and political structure which he saw in the society about him. Such a man could not be a "great psychologist" in the sense in which India and the West produced "great psychologists." He could, however, be a great systematizer of shrewd observations and practical rules for sound living in an ordered society; he could put into vivid, clear, and practical language the principles which all thoughtful men accepted. One might allow oneself a parallel here with Rome rather than with Greece: The emperor Marcus Aurelius was not capable of *creating* a philosophy of life, but he was capable of taking the current Stoic philosophy and stating it in a language acceptable to the thoughtful statesmen and law-givers of an empire which had to give first attention to stability, self-control, and the safeguarding of traditional values. Confucius must be judged in terms of the adequacy, clarity, and consistency of his definition of the good life, and in these terms, he comes off very well indeed.

### Aphorisms of Confucius*

THE MEASURE OF MAN IS MAN

Confucius said, "To one who loves to live according to the principles of true manhood without external inducements and who hates all that is contrary to the principles of true manhood without external threats of punishments, all mankind seems but like one man only. Therefore the superior man discusses all questions of conduct on the basis of himself as the standard, and then sets rules for the common people to follow."

Confucius said, "True manhood requires a great capacity and the road thereto is difficult to reach. You cannot lift it by your hands and you cannot reach it by walking on foot. He who approaches it to a greater degree than others may already be called 'a true man.' Now is it not a difficult thing for a man to try to reach this standard by sheer effort? Therefore, if the gentleman measures men by the standard of the absolute standard of righteousness, then it is difficult to be a real man. But if he measures men by the standard of man, then the better people will have some standard to go by."

* Trans. by Lin Yutang, from Lin Yutang (ed.), *The Wisdom of India and China* (New York: Random House, 1942). Copyright 1942 by Random House, Inc. Reprinted by permission.

Confucius said, "To a man who feels down in his heart that he is happy and natural while acting according to the principles of true manhood, all mankind seems like but one man." (What is true of the feelings of one person will serve as the standard of feelings for all people.)

Tse-kung asked, "If there is a man here who is a benefactor of mankind and can help the masses, would you call him a true man?" "Why, such a person is not only a true man," said Confucius, "he is a Sage. Even the Emperors Yao and Hsun would fall short of such a standard. Now a true man, wishing to establish his own character, also tries to establish the character of others, and wishing to succeed himself, tries also to help others to succeed. To know how to make the approach from one's neighbors (or from the facts of common, everyday life) is the method of formula for achieving true manhood."

Confucius said, "Is the standard of true manhood so far away, after all? When I want true manhood, there it is right by me."

## THE GOLDEN RULE

Chung Kung asked about true manhood, and Confucius replied, "When the true man appears abroad, he feels as if he were receiving distinguished people, and when ruling over the people, he feels as if he were worshipping God. What he does not want done unto himself, he does not do unto others. And so both in the state and in the home, people are satisfied."

Tse-kung said, "What I do not want others to do unto me, I do not want to do unto them." Confucius said, "Ah Sze, you cannot do it."

Confucius said, "Ah Ts'an, there is a central principle that runs through all my teachings." "Yes," said Tseng-tse. When Confucius left, the disciples asked Tseng-tse what he meant, and Tseng-tse replied, "It is just the principle of reciprocity."

Tse-kung asked, "Is there one single word that can serve as a principle of conduct for life?" Confucius replied, "Perhaps the word 'reciprocity' will do. Do not do unto others what you do not want others to do unto you."

## The Golden Mean of Tse-sze* †

### THE CENTRAL HARMONY

(I) What is God-given is what we call human nature. To fulfill the law of our human nature is what we call the moral law. The cultivation of the moral law is what we call culture.

---

* A grandson of Confucius.

† Trans. by Ku Hung-ming, from Lin Yutang (ed.), *The Wisdom of India and China* (New York: Random House, 1942). Copyright 1942 by Random House, Inc. Reprinted by permission.

The moral law is a law from whose operation we cannot for one instant in our existence escape. A law from which we may escape is not the moral law. Wherefore it is that the moral man (or the superior man) watches diligently over what his eyes cannot see and is in fear and awe of what his ears cannot hear.

There is nothing more evident than that which cannot be seen by the eyes and nothing more palpable than that which cannot be perceived by the senses. Wherefore the moral man watches diligently over his secret thoughts.

When the passions, such as joy, anger, grief, and pleasure have not awakened, that is our *central* self, or moral being (*chung*). When these passions awaken and each and all attain due measure and degree, that is *harmony*, or the moral order (*ho*). Our central self or moral being is the great basis of existence, and *harmony* or moral order is the universal law in the world.

When our true central self and harmony are realized, the universe then becomes a cosmos and all things attain their full growth and development.

THE GOLDEN MEAN

(II) Confucius remarked: "The life of the moral man is an exemplification of the universal moral order. The life of the vulgar person, on the other hand, is a contradiction of the universal moral order.

"The moral man's life is an exemplification of the universal order, because he is a moral person who unceasingly cultivates his true self or moral being. The vulgar person's life is a contradiction of the universal order, because he is a vulgar person who in his heart has no regard for, or fear of, the moral law."

(III) Confucius remarked: "To find the central clue to our moral being which unites us to the universal order, that indeed is the highest human attainment. For a long time, people have seldom been capable of it."

(IV) Confucius remarked: "I know now why the moral life is not practiced. The wise mistake moral law for something higher than what it really is; and the foolish do not know enough what moral law really is. I know now why the moral law is not understood. The noble natures want to live too high, high above their moral ordinary self; and ignoble natures do not live high enough, i.e., not up to their moral ordinary true self. There is no one who does not eat and drink. But few there are who really know flavor."

(V) Confucius remarked: "There is in the world now really no more social order at all."

(VII) Confucius remarked: "Men all say 'I am wise'; but when

149

driven forward and taken in a net, a trap, or a pitfall, there is not one who knows how to find a way of escape. Men all say, 'I am wise'; but in finding the true central clue and balance in their moral being (i.e., their normal, ordinary, true self), they are not able to keep it for a round month."

(VIII) Confucius remarked of his favorite disciple, Yen Huei: "Huei was a man who all his life sought the central clue in his moral being, and when he got hold of one thing that was good, he embraced it with all his might and never lost it again."

(IX) Confucius remarked: "A man may be able to put a country in order, be able to spurn the honors and emoluments of office, be able to trample upon bare, naked weapons; with all that he is still not able to find the central clue in his moral being."

(X) Tse-lu asked what constituted strength of character.

Confucius said: "Do you mean strength of character of the people of the southern countries or force of character of the people of the northern countries; or do you mean strength of character of your type? To be patient and gentle, ready to teach, returning not evil for evil; that is the strength of character of the people of the southern countries. It is the ideal place for the moral man. To lie under arms and meet death without regret; that is the strength of character of the people of the northern countries. It is the ideal of brave men of your type. Wherefore the man with the true strength of moral character is one who is gentle, yet firm. How unflinching is his strength! When there is moral social order in the country, if he enters public life he does not change from what he was when in retirement. When there is no moral social order in the country, he is content unto death. How unflinching is his strength!"

(XI) Confucius remarked: "There are men who seek for the abstruse and strange and live a singular life in order that they may leave a name to posterity. This is what I never would do. There are again good men who try to live in conformity with the moral law, but who, when they have gone halfway, throw it up. I never could give it up. Lastly, there are truly moral men who unconsciously live a life in entire harmony with the universal moral order and who live unknown to the world and unnoticed of men without any concern. It is only men of holy, divine natures who are capable of this."

MORAL LAW EVERYWHERE

(XII) The moral law is to be found everywhere, and yet it is a secret. The simple intelligence of ordinary men and women of the people may understand something of the moral law; but in its utmost reaches there is something which even the wisest and holiest of men cannot understand. The ignoble natures of ordinary men and women of the people may be able to carry out the moral law; but in its utmost reaches even the wisest and holiest of men cannot live up to it.

Great as the Universe is, man is yet not always satisfied with it. For there is nothing so great but the mind of the moral men can conceive of something still greater which nothing in the world can hold. There is nothing so small but the mind of the moral man can conceive of something still smaller which nothing in the world can split.

The Book of Songs says: "The hawk soars to the heavens above and fishes dive to the depths below." That is to say, there is no place in the highest heavens above nor in the deepest waters below where the moral law is not to be found. The moral man finds the moral law beginning in the relation between man and woman; but ending in the vast reaches of the universe.

(XVI) Confucius remarked: "The power of spiritual forces in the Universe—how active it is everywhere! Invisible to the eyes, and impalpable to the senses, it is inherent in all things, and nothing can escape its operation."

It is the fact that there are these forces which makes men in all countries fast and purify themselves, and with solemnity of dress institute services of sacrifice and religious worship. Like the rush of mighty waters, the presence of unseen Powers is felt; sometimes above us, sometimes around us.

In the Book of Songs it is said:

> The presence of the Spirit:
> It cannot be surmised,
> How may it be ignored!

Such is the evidence of things invisible that it is impossible to doubt the spiritual nature of man.

### THE HUMANISTIC STANDARD

(XIII) Confucius said: "Truth does not depart from human nature. If what is regarded as truth departs from human nature, it may not be regarded as truth. The Book of Songs says: 'In hewing an axe handle, the pattern is not far off.' Thus, when we take an axe handle in our hand to hew another axe handle and glance from one to the other, some still think the pattern is far off. Wherefore the moral man in dealing with men appeals to the common human nature and changes the manner of their lives and nothing more.

**151**

"When a man carries out the principles of conscientiousness and reciprocity he is not far from the moral law. What you do not wish others should do unto you, do not do unto them.

"There are four things in the moral life of a man, not one of which I have been able to carry out in my life. To serve my father as I would expect my son to serve me: that I have not been able to do. To serve my sovereign as I would expect a minister under me to serve me: that I have not been able to do. To act toward my elder brothers as I would expect my younger brother to act toward me: that I have not been able to do. To be the first to behave toward friends as I would expect them to behave toward me: that I have not been able to do.

"In the discharge of the ordinary duties of life and in the exercise of care in ordinary conversation, whenever there is shortcoming, never fail to strive for improvement, and when there is much to be said, always say less than what is necessary; words having respect to actions and actions having respect to words. Is it not just this thorough genuineness and absence of pretense which characterizes the moral man?"

(XV) The moral life of man may be likened to traveling to a distant place: one must start from the nearest stage. It may also be likened to ascending a height: one must begin from the lowest step. The Book of Songs says:

> When wives and children and their sires are one,
> 'Tis like the harp and lute in unison.
> When brothers live in concord and at peace
> The strain of harmony shall never cease.
> The lamp of happy union lights the home,
> And bright days follow when the children come.

Confucius, commenting on the above, remarked: "In such a state of things what more satisfaction can parents have?"

(XIV) The moral man conforms himself to his life circumstances; he does not desire anything outside of his position. Finding himself in a position of wealth and honor, he lives as becomes one living in a position of wealth and honor. Finding himself in a position of poverty and humble circumstances, he lives as becomes one living in a position of poverty and humble circumstances. Finding himself in uncivilized countries, he lives as becomes one living in uncivilized countries. Finding himself in circumstances of danger and difficulty, he acts according to what is required of a man under such circumstances. In one word, the moral man can find himself in no situation in life in which he is not master of himself.

In a high position he does not domineer over his subordinates. In a subordinate position he does not court the favors of his superiors. He puts in order his own personal conduct and seeks nothing from others; hence he has no complaint to make. He complains not against God, nor rails against men.

Thus it is that the moral man lives out the even tenor of his life calmly waiting for the appointment of God, whereas the vulgar person takes to dangerous courses, expecting the uncertain chances of luck.

Confucius remarked: "In the practice of archery we have something resembling the principle in a moral man's life. When the archer misses the center of the target, he turns round and seeks for the cause of his failure within himself."

CERTAIN MODELS

(VI) Confucius remarked: "There was the Emperor Shun. He was perhaps what may be considered a truly great intellect. Shun had a natural curiosity of mind and he loved to inquire into ordinary conversation. He ignored the bad (words?) and broadcast the good. Taking two extreme counsels, he took the mean between them and applied them in dealings with his people. This was the characteristic of Shun's great intellect."

(XVII) Confucius remarked: "The Emperor Shun might perhaps be considered in the highest sense of the word a pious man. In moral qualities he was a saint. In dignity of office he was the ruler of the empire. In wealth all that the wide world contained belonged to him. After his death his spirit was sacrificed to in the ancestral temple, and his children and grandchildren preserved the sacrifice for long generations.

"Thus it is that he who possesses great moral qualities will certainly attain to corresponding high position, to corresponding great prosperity, to corresponding great name, to corresponding great age.

"For God in giving life to all created things is surely bountiful to them according to their qualities. Hence the tree that is full of life, He fosters and sustains, while that which is ready to fall He cuts off and destroys.

The Book of Songs says:

> That great and noble Prince displayed
> The sense of right in all he wrought;
> The spirit of his wisdom swayed
> Peasant and peer; the crowd, the court.

"Being true to oneself is the law of God. Try to be true to oneself is the law of man.

"He who is naturally true to himself is one who, without effort, hits upon what is right, and without thinking understands what he wants to know, whose life is easily and naturally in harmony with the moral law. Such a one is what we call a saint or a man of divine nature. He who learns to be his true self is one who finds out what is good and holds fast to it.

"In order to learn to be one's true self, it is necessary to obtain a wide and extensive knowledge of what has been said and done in the world; critically to inquire into it; carefully to ponder over it; clearly to sift it; and earnestly to carry it out.

"It matters not what you learn; but when you once learn a thing, you must never give it up until you have mastered it. It matters not what you inquire into, but when you inquire into a thing, you must never give it up until you have thoroughly understood it. It matters not what you try to think out, but when you once try to think out a thing you must never give it up until you have got what you want. It matters not what you try to sift out, but when you once try to sift out a thing, you must never give it up until you have sifted it out clearly and distinctly. It matters not what you try to carry out, but when you once try to carry out a thing you must never give it up until you have done it thoroughly and well. If another man succeed by one effort, you will use a hundred efforts. If another man succeed by ten efforts, you will use a thousand efforts.

"Let a man really proceed in this manner, and, though dull, he will surely become intelligent; though weak, he will surely become strong."

(XXI) To arrive at understanding from being one's true self is called nature, and to arrive at being one's true self from understanding is called culture. He who is his true self has thereby understanding, and he who has understanding finds thereby his true self.

Therefore every system of moral laws must be based upon the man's own consciousness, verified by the common experience of mankind, tested by due sanction of historical experience and found without error, applied to the operations and processes of nature in the physical universe and found to be without contradiction, laid before the gods without question or fear, and able to wait a hundred generations and have it confirmed without a doubt by a Sage of posterity. The fact that he is able to confront the spiritual powers of the universe without any fear shows that he understands the laws of God. The fact that he is prepared to wait a hundred generations for confirmation from the Sage of posterity without any misgiving shows that he understands the laws of man.

Wherefore it is that it is true of the really great moral man that every move he makes becomes an example for generations; every act he does becomes a model for generations and every word he utters becomes a guide for generations.

## *TAOISM*

Many have found in Tao, "the Way," the central philosophical idea of ancient China. The great poem Tao Teh Ching (compare page 145), which expresses it, is at once intensely personal and intensely mystical. One feels, as one scans its rhythms, the gentle whimsies as well as the eternal verities of the man who wanted to say all that ever can be put into words, and knew that nothing really profound could be put into words—a man of infinite strength and of infinite gentleness. It is not surprising that Confucius, practical and prophetic teacher, found it hard to understand Lao-tzu, for none of us, while gently resonating to the receptive and practical wisdom of his words, can ever really say that he understands him. The two great teachers, contemporaries in the world of five hundred years before Christ, represent on the one hand the solid and practical, and on the other hand the intuitive and the mystical; yet both belong to the same world. If there be a paradox here, consider what Dr. Hsu has said about the fundamental practicality of the mystical words of Lao-tzu. It is not our duty here to try to weigh Lao-tzu in the balance. Rather, we hope to help the reader to see the deep sense in which he is both mystical and practical.

### *Confucius and Lao-tzu**

Confucius had lived to the age of fifty-one without hearing Tao, when he went south to P'ei, to see Lao-tzu.

Lao-tzu said, "So you have come, sir, have you? I hear you are considered a wise man up north. Have you got Tao?"

"Not yet," answered Confucius.

"In what direction," asked Lao-tzu, "have you sought for it?"

"I sought it for five years," replied Confucius, "in the science of numbers, but did not succeed."

"And then? . . ." continued Lao-tzu.

"Then," said Confucius, "I spent twelve years seeking for it in the doctrine of the Yin and Yang, also without success."

"Just so," rejoined Lao-tzu. "Were Tao something which could be presented, there is no man but would present it to his sovereign, or

---

* From Lewis Browne (ed.), *The World's Great Scriptures* (New York: The Macmillan Company, 1961). Copyright 1946 by Lewis Browne. Reprinted by permission of the Macmillan Company.

to his parents. Could it be imparted or given, there is no man but would impart it to his brother or give it to his child. But this is impossible, for the following reason. Unless there is a suitable endowment within, Tao will not abide. Unless there is outward correctness, Tao will not operate. The external being unfitted for the impression of the internal, the true sage does not seek to imprint. The internal being unfitted for the reception of the external, the true sage does not seek to receive.

"Reputation is public property; you may not appropriate it in excess. Charity and duty to one's neighbor are as caravanserais established by wise rulers of old; you may stop there one night, but not for long, or you will incur reproach.

"The perfect men of old took their road through charity, stopping a night with duty to their neighbor, on their way to ramble in transcendental space. Feeding on the produce of non-cultivation, and establishing themselves in the domain of no obligations, they enjoyed their transcendental inaction. Their food was ready to hand; and being under no obligations to others, they did not put anyone under obligation to themselves. The ancients called this the outward visible sign of an inward and spiritual grace.

"Those who make wealth their all in all, cannot bear loss of money. Those who make distinction their all in all, cannot bear loss of fame. Those who affect power will not place authority in the hands of others. Anxious while holding, distressed if losing, yet never taking warning from the past and seeing the folly of their pursuit—such men are the accursed of God.

"Resentment, gratitude, taking, giving, censure of self, instruction of others, power of life and death—these eight are the instruments of right; but only he who can adapt himself to the vicissitudes of fortune, without being carried away, is fit to use them. Such a one is an upright man among the upright. And he whose heart is not so constituted—the door of divine intelligence is not yet opened for him."

Confucius visited Lao-tzu, and spoke of charity and duty to one's neighbor.

Lao-tzu said, "The chaff from winnowing will blind a man's eyes so that he cannot tell the points of the compass. Mosquitoes will keep a man awake all night with their biting. And just in the same way this talk of charity and duty to one's neighbor drives me nearly crazy. Sir! strive to keep the world to its own original simplicity. And as the wind bloweth where it listeth, so let virtue establish itself. Wherefore such undue energy, as though searching for a fugitive with a big drum?

"The snow goose is white without a daily bath. The raven is black without daily coloring itself. The original simplicity of black and of white is beyond the reach of argument. The vista of fame and reputation is not worthy of enlargement. When the pond dries up and the fishes are left upon dry ground, to moisten them with the breath or to damp them with a little spittle is not to be compared with leaving them in the first instance in their native rivers and lakes."

On returning from this visit to Lao-tzu, Confucius did not speak for three days. A disciple asked him, saying, "Master, when you saw Lao-tzu, in what direction did you admonish him?"

"I saw a dragon," replied Confucius, "—a dragon which by convergence showed a body, by radiation became color, and riding upon the clouds of heaven, nourished the two principles of creation. My mouth was agape: I could not shut it. How then do you think I was going to admonish Lao-tzu?"

We turn now to Lao-tzu's philosophy, "The Way," the Tao Teh Ching.*

## 1

Tao can be talked about, but not the Eternal Tao.
Names can be named, but not the Eternal Name.

As the origin of heaven-and-earth, it is nameless:
As "the Mother" of all things, it is nameable.

So, as ever hidden, we should look at its inner essence:
As always manifest, we should look at its outer aspects.

These two flow from the same source, though differently named;
And both are called mysteries.

The Mystery of mysteries is the Door of all essence.

## 2

When all the world recognizes beauty as beauty, this in itself is ugliness.
When all the world recognizes good as good, this in itself is evil.

Indeed, the hidden and the manifest give birth to each other.
Difficult and easy complement each other.
Long and short exhibit each other.
High and low set measure to each other.
Voice and sound harmonize each other.
Back and front follow each other.

* From Lao Tzu, *Tao Teh Ching*, Paul Sih (ed.), Asian Institute Translations, No. 1 (New York: St. John's University Press, 1961). Copyright 1961 by St. John's University.

Therefore, the Sage manages his affairs without ado,
And spreads his teaching without talking.
He denies nothing to the teeming things.
He rears them, but lays no claim to them.
He does his work, but sets no store by it.
He accomplishes his task, but does not dwell upon it.

And yet it is just because he does not dwell on it
That nobody can ever take it away from him.

### 3

By not exalting the talented you will cause the people to cease from
      rivalry and contention.
By not prizing goods hard to get, you will cause the people to cease from
      robbing and stealing.
By not displaying what is desirable, you will cause the people's hearts
      to remain undisturbed.

Therefore, the Sage's way of governing begins by

        Emptying the heart of desires,
        Filling the belly with food,
        Weakening the ambitions,
        Toughening the bones.

In this way he will cause the people to remain without knowledge and
      without desire, and prevent the knowing ones from any ado.
Practice Non-Ado, and everything will be in order.

### 6

The Spirit of the Fountain dies not.
It is called the Mysterious Feminine.
The Doorway of the Mysterious Feminine*
Is called the Root of Heaven-and-Earth.

Lingering like gossamer, it has only a hint of existence;
And yet when you draw upon it, it is inexhaustible.

### 7

Heaven lasts long, and Earth abides.
What is the secret of their durability?
Is it not because they do not live for themselves
That they can live so long?

   * Goethe: "the eternal feminine draws us on" (closing lines of *Faust*).

Therefore, the Sage wants to remain behind,
But finds himself at the head of others;
Reckons himself out,
But finds himself safe and secure.
Is it not because he is selfless
That his Self is realized?

### 8

The highest form of goodness is like water.
Water knows how to benefit all things without striving with them.
It stays in places loathed by all men.
Therefore, it comes near the Tao.

In choosing your dwelling, know how to keep to the ground.
In cultivating your mind, know how to dive in the hidden deeps.
In dealing with others, know how to be gentle and kind.
In speaking, know how to keep your words.
In governing, know how to maintain order.
In transacting business, know how to be efficient.
In making a move, know how to choose the right moment.

If you do not strive with others,
You will be free from blame.

### 9

As for holding to fullness,
Far better were it to stop in time!

Keep on beating and sharpening a sword,
And the edge cannot be preserved for long.

Fill your house with gold and jade,
And it can no longer be guarded.

Set store by your riches and honor,
And you will only reap a crop of calamities.

Here is the Way of Heaven:
*When you have done your work, retire!*

### 10

In keeping the spirit and the vital soul together,
Are you able to maintain their perfect harmony?
In gathering your vital energy to attain suppleness,
Have you reached the state of a newborn babe?

In washing and clearing your inner vision,
Have you purified it of all dross?
In loving your people and governing your state,
Are you able to dispense with cleverness?
In the opening and shutting of heaven's gate,
Are you able to play the feminine part?
Enlightened and seeing far into all directions,
Can you at the same time remain detached and non-active?

We think there is enough here to show that Lao-tzu is *both* practical and mystical. Some modern authors, however, emphasize the mystical. For example, R. B. Blakney*:

Wherever the great mysticism has come, it has offered to replace popular or local religion with a new and universal allegiance. Folk beliefs about gods and spirits give place to a metaphysic of the utmost generality for those who can rise to it. The mystic's passion is satisfied only with the sense of the Ultimate Reality, the God, Godhead or Godness that is back of the world of mind and nature. What is the Ultimate like? And what has it to do with man? The mystic report is that:

*Reality, however designated, is One; it is an all-embracing unity from which nothing can be separated.* "Hear, O Israel; The Lord our God is one Lord." (Palestine, seventh century B.C.) "So I say that likeness born of the One, leads the soul to God, for he is One, unbegotten unity, and of this we have clear evidence." (Eckhart, Germany, 1300.) "Behold but One in all things; it is the second that leads you astray." (Kabir, India, 1500.) "Something there is, whose veiled creation was before the earth or sky began to be; so silent, so aloof and so alone, it changes not nor fails but touches all." (Tao Teh Ching, 25.)

*IT, the Ultimate, is nameless, indescribable, beyond telling: and therefore anything said about it is faulty:* " . . . What is his name? . . . And God said to Moses, I AM WHAT I AM . . . say unto the children of Israel, I AM hath sent me unto you." (Exodus 3:14.) "Describe it as form yet unformed, as shape that is still without shape; or say it is vagueness confused: one meets it and it has no front; one follows and there is no rear." (14.) "IT cannot be defined by word or idea; as the Scripture says, it is the One 'before whom words recoil.'" (Shankara-charya, India, 800.) "It is God's nature to be without a nature. To think of his goodness, or wisdom, or power is to hide the essence of him, to

* From *The Way of Life*, a new translation of the Tao Teh Ching by R. B. Blakney (New York: New American Library, 1955), pp. 29–31.

obscure it with thoughts about him. . . . Who is Jesus? He has no name."
(Eckhart.)

*Within the self, IT is to be found and there it is identical with Reality
in the external world:* "So God created man in his own image, in the
image of God created he him. . . ." (Genesis 1:27.) ". . . the Father
is in me and I in the Father. . . . I and the Father are one." (John
10:38, 30.) "As sure as the Father, so single in nature, begets his Son,
he begets him in the spirit's inmost recess, and that is the inner world.
Here, the core of God is also my core; and the core of my soul, the core
of God's. . . ." (Eckhart.) "Here likewise in this body of yours, my
son, you do not perceive the True; but there in fact it is. In that which
is the subtle essence, all that exists has its self. That is the True, that
is the Self, and thou, Svetaketu, art That." (Chandogya Upanishad,
India.) "I went from God to God, until they cried from me in me, 'O
thou I!' " (Bayazid, Persia, 847.) "The world may be known without
leaving the house; God's way may be seen apart from the windows."

*IT can be known, not discursively, but by acquaintance, and this ac-
quaintance is the point of all living:* "This is Life Eternal, to know you,
the one true God. . . ." (John 17:3.) "Know that when you learn to lose
yourself, you will reach the Beloved. There is no other secret to be
learnt, and more than this is not known to me." (Ansari of Herat.)
"Where is this God? In eternity. Just as a man who is in hiding clears
his throat and reveals his whereabouts, so it is with God. Nobody could
ever find God. He has to discover himself. . . . But when one takes God
as he is divine . . . he will be like one athirst; he cannot help drinking
even though he thinks of other things . . . the idea of the Drink will not
depart as long as the thirst endures." (Eckhart.) "If you work by the
Way, you will be of the Way; if you work through its power, you will
be given the power; abandon either one and both abandon you."

*Reality is disclosed only to those who meet its conditions and the
conditions are primarily moral:* "Blessed are the pure in heart: for they
shall see God." (Matthew 5:8.) "The more a man regards everything
as divine—more divine than it is of itself—the more God will be pleased
with him. To be sure, this requires effort and love, a careful cultivation
of the spiritual life, and a watchful, honest, active oversight of all one's
mental attitudes toward things and people. It is not to be learned by
world-flight, running away from things, turning solitary and going apart
from the world. Rather, one must learn an inner solitude. He must learn
to penetrate things and find God there, to get a strong impression of
God fixed firmly in his mind." (Eckhart.) "When the heart weeps for
what it has lost, the spirit laughs for what it has found." (Sufi.) "With
the lamp of word and discrimination one must go beyond word and
discrimination and enter upon the path of realization." (Lankavantara

Sutra.) "The student learns by daily increment. The Way is gained by daily loss, loss upon loss until at last comes rest."

This outline of illustrative material taken from the writings of mystics of diverse lands and times is, of course, intended only to suggest the uplands of the spirit touched by those who wrote the Tao Teh Ching. The Chinese purview of those uplands was somewhat bleaker than that of many other great mystic writings, perhaps because the difficulties of calligraphy in ancient China demanded severe terseness. It may also have been due to certain esoteric requirements put upon its poems: they suggest that aside from the king, others for whom they were intended had already been disciplined through long renunciation to understand their cryptic references. Clearly, the Way had to be learned; it could not be taught.

On mature acquaintance, however, what at first appears as bleakness in these poems becomes a powerful simplicity of conception and a sureness of touch in which there is no waste motion. What can be said of the Way is said with great economy of language, an economy like that of lightning in the night. The Virtue or power of the Way is illustrated by an assertion of the consequences of letting the Way be the way by which the affairs of men are dispatched. This is the radical element in the poems. Reality, the Way, is taken with great seriousness. Why should men interfere with its operations? It would be much better if, from the king down to the least peasant, the wills of men should be held subject to the Way, so that like all else in the universe, men should become its perfect instruments. Let be! Then the mighty Way will act through you and its power will accomplish what you cannot do of your own volition.

Two centuries after Lao-tzu came Chuang-tzu, who, instead of poetic intuition, used orderly philosophic discussion in the service of "the Way." Vigorously anti-Confucianist, he looked to philosophy as a means of discovering and accepting the world as it is, free from rules, admonitions, and coercions. A good life withdraws from excitement and challenge. When ambassadors from the state of Ch'u invited him at the riverside to become prime minister, he went on with his fishing. The world is one and indivisible. It is good. It is beautiful. And it is there to be lived in. One might think of Stoicism; but Stoicism is severe, and there is only charm, no severity, in the Taoism of Chuang-tzu.

The following discussion of Chuang-tzu was written by our consultant, Dr. Hsu.

### Chuang-tzu

Together with the book of Lieh-tzu and the Tao Teh Ching by Lao-tzu, Chuang-tzu's dialogues rank as one of the earlier classics of Taoism. The antiquity of Lieh-tzu has been questioned and the clarity of the Tao Teh Ching is often doubted. Only Chuang-tzu's book provides us with a well-developed philosophy.

Chuang-tzu's philosophy might be taken as an effort of reconciliation between what William James describes as the "tough-minded" and "tender-minded" philosophies—those which reduce everything to a materialistic basic and those which elevate everything to a spiritualistic plane. But in fact Chuang-tzu was not fighting any such battle. He was not consciously effecting any reconciliation. He believes in the all-pervading and all-embracing reality which is all men, gods, and things: "complete, all-embracing, and the whole." There is not a single thing without Tao. "The totality of the spontaneity of all things is Tao." He also believes in Teh, which is related to Tao as water in river or lake is to water in general. "The total spontaneity of all things is Tao. The spontaneity that an individual thing receives from Tao is Teh."

The essence of Chuang-tzu's philosophy is letting-alone. But letting-alone does not mean absence of action. It means non-interference with the courses of others and of the self. "Act according to your will within the limit of your nature, but have nothing to do with what is beyond it." Let the world alone, to work out its own salvation, whether it is in peace or disorder. Don't try to impose upon the world your own standards.

As a personal philosophy, Chuang-tzu enjoins the individual to be above both praise and detraction, to change with time and to insist upon nothing. The "perfect man is both tranquil and active," for "in the perfect man tranquility and activity unite." The perfect man enjoys the absolute freedom of not being dependent upon anything: "he has transcended all distinctions." This in essence is identical with the philosophy of Spinoza, who asks the individual to lose himself in nature or God and there find his salvation. However, in spite of such a philosophy, and the extent to which Confucius subscribed to it, the Chinese way of life differs drastically from that of the West and of the Hindu.

There is a strong emphasis in Chuang-tzu upon trends in the universe and upon the natural pattern of functioning for living organisms. One bird, the *p'eng*, can fly in a high place, the quail in the low; one vegetable, the *ta'chun*, can live for a long time, the mushroom for only a short while. "All these capacities are normal, not made or learned . . . thus they are normal."

This concept of following the natural way, or the nature of things, was expressed in different ways and seems to be reflected in many other specific attitudes and points of view: "To ride upon the transformation of the six elements is to make an excursion upon the road of change and evolution. If one is going on like this, when can one reach the end?"

The solution to the problem presented to the individual by the facts of change is not to be dependent upon any one thing. "If one has to depend upon something, one cannot be happy, unless one gets hold of the thing which he depends upon. . . . Only he who ignores the distinction between things and follows a great evolution can be really independent and always free." The implication here seems to be that danger is attached to specific things which may be lost or changed, that only by ignoring the distinction between things and being ready to accept newness is it possible to be free.

Related to this is the idea that "the perfect man has no self; the spiritual man has no achievement." We are dealing here with the problem of egocentricity and aggrandizement: "The independent man . . . forgets his own self and ignores all differences." This is extended to a concept that if the distinction itself is insisted upon, this brings early death. "Those whose happiness is attached within the finites feel that they certainly have limitations. . . . If things enjoy themselves only in their finite spheres, their enjoyment must also be finite. . . . If one enjoyed only in life he would suffer in death. . . . If one enjoyed only in power, he would suffer loss of it." The "independent man" transcends the finite. He thus becomes infinite and so is his happiness. "The perfect man has no self" (that is, no bounded, self-centered attitude) because he has transcended the finite and identified himself with the universe. "The spiritual man has no achievement" because he follows the nature of things and lets everything enjoy itself.

This seems to imply that by accepting the universe as a whole and responding to different aspects of it, participating in and accepting change, one transcends the finite and avoids the suffering which is caused by attachment to achievement and desire for specific things.

Psychological conditions are accepted with the statement that it is not possible to tell how they arise: "Joy and anger, sorrow and pleasure, anxiety and regret, fickleness and determination, vehemence and indolence, indulgence and extravagance: these come like music sounding from an empty tube, or mushrooms springing out of warmth and moisture. Daily and nightly they alternate within us, but we cannot tell whence they spring. Can we find out in a moment how they are produced?"

Some of the descriptions of differences of different minds seem to imply the equal value of different styles. "The mind of some flies forth like a javelin, the arbiter of right and wrong. The mind of others remains firm, like a solemn covenanter, the guardian of rights secured." That is, different styles of approach can be utilized in different social roles. It may not be going too far to suggest that here the Chinese author anticipates the thought of Erik Erikson, who deals both with the social milieu in which development takes place and with the social niche in which a given character can find and fulfill himself.

A more extreme example of the conception of equal value of different aspects of human functioning appears in the formulation: "The hundred parts of the human body, with its nine openings, and six viscera, all are complete in their places. Which shall I prefer? Do you like them all equally or do you like some more than others?" Here the most studied scientific objectivity could not transcend Victorian and Biblical prejudices against the "lower" aspects of our nature and of our bodies.

In further discussions it is implied that there is no absolute criterion of value in respect to right and wrong either; all distinctions of right and wrong are due to opinions. What is affirmed by one may be denied by another.

"From the point of view of things, everything values itself and considers others worthless . . . if we say a thing is great because it is considered great by something, then everything is great . . . if we say a thing is right because it is considered right by something, then everything is right . . . any word can be the predicate of anything."

This relativity is restated more basically: "Everything is 'that' (another thing's other); everything is 'this' (its own self). Things do not know that they are another's 'that'; they only know that they are 'this.' The 'that' and the 'this' produce each other."

Similarly, because of the right there is a wrong, because of the wrong there is the right.

But when one looks at things from the (disinterested or objective) "point of view of nature" the "this" is also "that." The "that" is also "this." The very essence of Tao is not to discriminate "that" and "this" as opposites.

This curious mixture of mysticism and whimsy appears in two little stories:

Once upon a time, I, Chuang-tzu, dreamt I was a butterfly, fluttering hither and thither, to all intents and purposes a butterfly. I was conscious only of following my fancies as a butterfly, and was unconscious of my individuality as a man. Suddenly, I awaked, and there I lay, myself again. Now I do not know whether I was then a man

165

dreaming I was a butterfly, or whether I am now a butterfly dreaming I am a man.*

Chuang-tzu and Hui-tzu were strolling one day on the bridge over the river Hao. Chuang-tzu said, "Look how the minnows dart hither and thither where they will. Such is the pleasure that fish enjoy." Hui-tzu said, "You are not a fish. How do you know what gives pleasure to fish?" Chuang-tzu said, "You are not I. How do you know that I do not know what gives pleasure to fish?" Hui-tzu said, "If because I am not you, I cannot know whether you know, then equally because you are not a fish, you cannot know what gives pleasure to fish. My argument still holds." Chuang-tzu said, "Let us go back to where we started. You asked me how I knew what gives pleasure to fish. But you already knew how I knew it when you asked me. You knew that I knew it by standing here on the bridge at Hao."**

The following selections from the writings of Chuang-tzu are from the translation by Yu-Lan Fung, a contemporary Chinese scholar.

### The Equality of Things and Opinions†

Nan Kuo Tzu Ch'i sat leaning on a table. He looked to heaven and breathed gently, seeming to be in a trance, and unconscious of his body.

Yen Ch'eng Tzu Yu, who was in attendance on him, said: "What is this? Can the body become thus like dry wood, and the mind like dead ashes? The man leaning on the table is not he who was here before."

"Yen," said Tzu Ch'i, "your question is very good. Just now, I lost myself, do you understand? You may have heard the music of man, but not the music of earth; you may have heard the music of earth, but not the music of heaven."

"I venture," said Tzu Yu, "to ask from you a general description of these."

"The breath of the universe," said Tzu Ch'i, "is called the wind. At times, it is inactive. When it is active, angry sounds come from every aperture. Have you not heard the growing roar? The imposing appearance of mountain forest, the apertures and cavities in huge trees many a span in girth: these are like nostrils, like mouth, like ears, like beam

* From Lewis Browne (ed.), *The World's Great Scriptures* (New York: The Macmillan Company, 1961), p. 332.

** From Arthur Waley, *Three Ways of Thought in Ancient China* (Garden City, N.Y.: Doubleday & Co., 1956), p. 7.

† From Yu-Lan Fung, *Chuang Tzu* (Shanghai: Commercial Press, Ltd., 1933).

sockets, like goblets, like mortars, like pools, like puddles. The wind goes rushing into them, making the sounds of rushing water, of whizzing arrows, of scolding, of breathing, of shouting, of crying, of deep wailing, of moaning agony. Some sounds are shrill, some deep. Gentle winds produce minor harmonies; violent winds, major ones. When the fierce gusts pass away, all the apertures are empty and still. Have you not seen the bending and quivering of the branches and leaves?"

"The music of earth," Tzu Yu said, "consists of sounds produced on the various apertures; the music of man, of sounds produced on pipes and flutes. I venture to ask of what consists the music of heaven."

"The winds as they blow," said Tzu Ch'i, "differ in thousands of ways, yet all are self-produced. Why should there be any other agency to excite them?"

Great knowledge is wide and comprehensive; small knowledge is partial and restricted. Great speech is rich and powerful; small speech is merely so much talk. When people sleep, there is confusion of soul; when awake, there is movement of body.

In the association of men with men, there are plotting and scheming; and daily there is striving of mind with mind. There are indecisions, concealments, and reservations. Small apprehensions cause restless distress, great apprehensions cause endless fear. The mind of some flies forth, like a javelin, the arbiter of right and wrong. The mind of others remains firm, like a solemn covenantor, the guardian of rights secured. The mind of some fails like decay in autumn and winter. The mind of others is sunk in sensuous pleasure and cannot come back. The mind of yet others has fixed habits like an old drain; it is near to death and cannot be restored to vigor. Joy and anger, sorrow and pleasure, anxiety and regret, fickleness and determination, vehemence and indolence, indulgence and extravagance: these come like music sounding from an empty tube, or mushrooms springing out of warmth and moisture. Daily and nightly they alternate within us, but we cannot tell whence they spring. Can we expect in a moment to find out how they are produced?

If there is no other, there will be no I. If there is no I, there will be none to make distinctions. This seems to be true. But what causes these varieties? It might seem as if there would be a real Lord, but there is no indication of His existence. One may believe that He exists, but we do not see His form. He may have reality, but no form. The hundred parts of the human body, with its nine openings, and six viscera, all are complete in their places. Which shall I prefer? Do you like them all equally? Or do you like some more than others? Are they all servants? Are these servants unable to control each other, but need another as ruler? Do they become rulers and servants in turn? Is there any true ruler other than themselves? Regarding these questions, whether we can obtain true answers or not, it matters but little to the reality of the ruler (if there is one).

Tao is obscured by partiality. Speech is obscured by eloquence. Therefore, there are the contentions between the Confucianists and the Mohists. Each one of these two schools affirms what the other denies, and denies what the other affirms. If we are to affirm what these two schools both deny, and to deny what they both affirm, there is nothing better than to use the light of reason.

Everything is "that" (another thing's other); everything is "this" (its own self). Things do not know that they are another's "that"; they only know that they are "this." The "that" and the "this" produce each other. Nevertheless, when there is life, there is death. When there is impossibility, there is possibility. Because of the right, there is the wrong. Because of the wrong, there is the right. On account of this fact, the sages do not take this way, but see things from the point of view of nature. The "this" is also "that." The "that" is also "this." According to "that," there is a system of right and wrong. According to this, there is also a system of right and wrong. Is there really a distinction between "that" and "this"? Is there really no distinction between "that" and "this"? Not to discriminate "that" and "this" as opposites is the very essence of Tao. Only the essence, an axis as it were, is the center of the circle responding to the endless changes. The right is an endless change. The wrong is also an endless change. Therefore, it is said that there is nothing better than to use the light of reason.

To wear out one's spirit and intelligence in order to unify things without knowing that they are already in agreement, is called "three in the morning." What is "three in the morning"? A keeper of monkeys once ordered concerning the monkey's rations of acorns that each monkey was to have three in the morning and four at night. But at this the monkeys were very angry. So the keeper said that they might have four in the morning, but three at night. With this arrangement, all monkeys were well pleased. . . .

"These points," said Chang Wu Tzu, "would have perplexed even the Yellow Emperor; how could Confucius be competent to understand them? Moreover, you are too hasty in forming your estimate. You see an egg, and immediately you expect to hear it crow. You look at the crossbow, and immediately you expect to have a roast dove before you. I shall speak a few words with you at random, and do you also listen to them at random? How does the sage sit by the sun and the moon, and hold the universe in his arm? He blends everything into a harmonious whole, rejects the confusion of distinctions, and ignores the differences in social rank. Men in general bustle about and toil; the sage is primitive and without knowledge. He blends together ten thousand years, and stops at the one, the whole, and the simple. All things are what they are, and spontaneously pursue their courses. How do I know that the love of life is not a delusion? How do I know that he who is afraid of death is not like a man who was away from his home when young and therefore has no intention to return? Li Chi was the daughter of the border warden of Ai. When the state of Chin first got her, she wept until the front part of her robe was drenched with tears. But when she came to the royal residence, shared with the king his luxurious couch and ate rich food, she regretted that she had wept. How do I know that the dead will not repent of their former craving for life? Those who dream of a banquet at night may in the next morning wail and weep. Those who dream of wailing and weeping may in the morning go out to hunt. When they dream, they do not know that they are dreaming. In their dream, they may even interpret dreams. Only when they are awake, they begin to know that they dreamed. By and by comes the great awakening, and then we shall find out that life itself is a great dream. All the while, the fools think that they are awake; that they know. With nice discriminations, they make distinctions between princes and grooms. How stupid! Confucius and you are both in a dream. When I say that you are in a dream, I am also in a dream. This saying is called a paradox. If after ten thousand ages we could once meet a great sage who knows how to explain it, it would be as if we meet him in a very short time.

"Suppose that you argue with me. If you beat me, instead of my beating you, are you necessarily right, and am I necessarily wrong? Or, if I beat you and not you me, am I necessarily right, and are you necessarily wrong? Is the one of us right and the other wrong? Or are both of us right and both of us wrong? Both of us cannot come to a mutual and common understanding, and others are all in the dark. Whom shall I ask to decide this dispute? I may ask someone who agrees with you; but since he agrees with you, how can he decide it? I may ask someone who agrees with me; but since he agrees with me, how can he decide it? I may ask someone who differs both from you and me; but since he differs both from you and me, how can he decide it? I may ask someone who agrees both with you and me; but since he agrees both with you and me, how can he decide it?"

## CHINESE CONCEPTIONS OF HUMAN NATURE

Wanting to go beyond the classical philosophers, and to attempt some generalization about Chinese ways of looking at human nature, we asked Dr. Francis Hsu to give us an example of the quality of Chinese thought as contrasted with Western thought. He emphasized the fundamental points that Chinese life is "situation-centered"; psychology is less "inward," more socially controlled; and herein lie many clues for the individuality of the kinds of psychology developed in China. The following passage is from Dr. Hsu's *Americans and Chinese*.*

In situation-centered China, sex is controlled more by external barriers than by internal restraints. Where the emphasis is on internal restraints, man is enjoined not only to avoid sinful action but also to eliminate sinful thoughts. But where control is exercised through external barriers, the individual needs merely to refrain from sexual expression in inappropriate situations.

Yet sex, being one of man's most fundamental urges, can hardly be eliminated from thought. Where attempts are made to eliminate it from consciousness by condemning it as bad, it merely takes refuge in the deeper layers of the mind, a condition Freud described as repression. Where the purpose is merely to regulate it, it is simply channeled into specific areas where one need not feel reserved. In the former approach, sex is a sin, regardless of circumstances. In the latter approach, sex is a natural urge of man, like eating, to which expression must be given in the right place and with the right parties.

Consequently, repression of sex is correlated with the generalized interest in it in American art, where sex appears very frequently but in the form of diffused suggestiveness. Compartmentalization of sex, on the other hand, causes it to be practically absent in general Chinese art, but concentrated without restraint in pornography. Except where sublimated, the desirability of sex increases in reverse ratio to its availability. Hence the emotional energy directed toward sex often outweighs that toward all other subjects of Western art while it is without comparable significance in Chinese art.

These basic contrasts in Chinese and American art are equally obvious in fiction. Chinese traditional novels as a rule concentrate upon what the characters do in their roles as emperors or common men, while American novels are much more concerned with what the characters *do, think,* and *feel* as individuals. There are, it is true, a few exceptions. But compare any Chinese novel which has remained popular throughout several centuries, such as *The Dream of the Red Chamber* or *All Men Are Brothers* with any widely read American novel, such as *The Grapes of Wrath* or *Elmer Gantry*. The absence in Chinese novels of introspec-

* London: The Cresset Press, 1955. Pp. 22–29.

tive excursions into the mind of a character is as pronounced as its relative abundance in American fiction. No traditional Chinese novelist told the whole story through the eyes of one character, which, however, is the principal technique of his American counterpart.

Among the works of Chinese fiction with which I am familiar, the deepest an author has penetrated into the workings of a character's mind was when a hero, confronted by an enemy, calculated that if he did a certain thing he would surely defeat or outwit the villain. The height of Chinese imaginative authorship seems to have been reached in *Western Journey*, a fantasy about an arduous trip to India by some Buddhist monks in the early seventh century A.D., and *Flowers in a Mirror*, a satire on human relations. Each is a sort of Chinese *Alice in Wonderland*.

No less revealing is the way love is handled in the two types of novels. In the American love story, the union of the hero and the heroine is usually the highest climax. Not infrequently an entire book deals with the pursuit of romance, the agony, the misunderstandings, the stumbling blocks that must be hurdled to reach that point where the two chief characters join each other and the book comes to an end. Even where love is not the question, as in *The Grapes of Wrath*, the American novel pulsates with strong emotions. The Chinese novel, even where it deals with romance, does so with a casualness or frankness that must be distasteful to the American reader. Sex union usually occurs early in the narrative. It is never the climax of the story. The central problem with which the balance of the novel is concerned is how the hero goes about marrying the heroine properly, with the rectifying wedding ceremony tediously described to the last detail. Mutual attraction between an individual man and woman is not enough. Their personal feelings are never more important than the sanctions of the group.

Furthermore, Chinese and American novelists described two very different kinds of romantic characters. With few exceptions, such as the warrior in *Captain from Castile* or the advertising executive in *The Hucksters*, an American hero and his lady are devoted to one another. This devotion often approaches worship. In particular, there is a tendency for the male characters to exalt their female idols in thought, words, or action.

A Chinese hero, on the other hand, pursues his female love object and revels in her beauty in the same way that he strives to gain control of other worldly goods, such as gold or jewelry. In *The Dream of the Red Chamber*, the hero, although never sexually united with the heroine, had illicit relations with a number of other women, including servant maids, while he was courting his ideal sweetheart. In addition, he also had erotic connections of varying degrees with a number of other women and homosexual relations with an actor. To American readers such a characterization probably befits that of a villain who would at least have to repent before becoming acceptable. But Chinese readers did not condemn this character. Lacking the individualized approach to sex, for many decades they have referred to him and his girl as great lovers.

These contrasts are not confined to but a handful of Chinese and American novels. One could go on almost indefinitely selecting examples from each group. Among the most popular Chinese novels that we may cite, in addition to those already mentioned, are *The Romance of the Three Kingdoms; The True Story of Chi Kung, the Mad Monk; The True Story of His Eminence Pao, the Wonderful Official; The Golden Lotus; Western Chamber;* and *Strange Stories from a Chinese Studio.* Any reader reasonably well acquainted with American fiction could prepare a similar and extensive list of novels that bear witness to the characteristics outlined previously. I think the reader would have a harder time finding American novels that do not possess the characteristics described here.

The diversity of subject matter in American and Chinese art—the near universal preoccupation with sex in American art and its segregation into a separate area in Chinese art—is to be found also in the content of American and Chinese fiction. Love stands out today, as at all times, as the most important theme of native American novels as well as of other Western novels which have enjoyed wide acceptance in America. There are, of course, a number of great European and American works, such as Thomas Mann's *The Magic Mountain,* Thomas Wolfe's *Look Homeward, Angel,* Feodor Dostoevski's *Crime and Punishment,* John Hersey's *A Bell for Adano,* Sinclair Lewis' *Babbitt* or *Main Street,* Nordoff and Hall's *Mutiny on the Bounty,* or Marjorie K. Rawlings' *The Yearling,* which have treated problems and situations that are not concerned essentially with the love theme. Too, there are famous authors such as Mark Twain, Herman Melville, and Jack London, whose productions have scarcely touched upon romance. I doubt, however, that anyone would choose to contend that the vast majority of Western novels do not deal with love.

The most popular Chinese novels have been primarily of two kinds. There are those which do not deal with the man-woman relationship at all, or only deal with it hurriedly as a minor matter; and there are pornographic novels which deal with little else.

*The Romance of the Three Kingdoms* is typical of the kind in which romance is merely a bypath along the main route of the story. This novel was based upon that period of China's history, about 200 A.D., when the country was divided among three warring factions. A synopsis of one brief section of this ten-volume work will indicate to what extent "romance" figures in Chinese novels of this variety.

At one point, the mastermind of the Eastern faction, the kingdom of Wu, designed the following strategy to destroy the head of the Western faction, the kingdom of Shu. The ruler of Shu was in need of a wife and the king of Wu had a beautiful sister. The story was spread that the king of Shu could marry this desirable woman if he came in person to the court of Wu. The strategist's idea was to kill the wife-seeker at a reception in his honor. The king of Shu, reluctant to accept the invitation, hesitated until his chief strategist advised him to go, accompanied by one lone warrior. What the Wu strategist failed to foresee was that the mother of the king of Wu insisted on attending the banquet to inspect her prospective son-in-law at close range. The elderly dowager at once developed a liking for the king of Shu. She decided that he was the right man for her daughter, and he was married to the beautiful woman with all the ceremonial pomp due a king, against the wishes of the Wu strategist.

Having lost the first round, the unruffled Wu strategist worked out a second plan. The king of Shu was showered with all kinds of gifts. A palace was built for him and his new bride. This palace was furnished with all manner of luxuries, staffed with a host of servants and beautiful girls, and here he was entertained lavishly. The Wu strategist's idea was that, having lived in such ease and comfort, the king of Shu would be unwilling to leave his gracious surroundings. He also counted on the princess persuading her husband to stay on indefinitely.

Unfortunately for the kingdom of Wu, the Shu strategist was one step ahead. He foresaw this latter eventuality and provided his warrior with a plan to deal with it. The upshot was that the Wu princess, far from wanting her husband to stay, insisted on going with her husband to the latter's kingdom. A second attempt at assassination was out of the question because this would widow the princess. The pair could not be separated because the princess remained loyal to her husband. Result: the king of Wu lost the battle of strategy as well as a beautiful sister. The outcome so humiliated the Wu strategist that he eventually died of anger, which, I suspect, was the popular American disease—high blood pressure.

There are only a few other romantic episodes in this novel. The ruler of the northern faction, the kingdom of Wei, was routed in battle because he neglected strategy while dallying with another man's wife. This was a typically Chinese piece of didacticism in which romance was no more than a vehicle for teaching duty's lesson—that the fulfillment of personal desires runs a poor second to many other considerations, and the failure to heed this injunction has disastrous consequences. In another instance, a great general, serving under the ruler of Shu, won a wide expanse of territory after refusing marital advances from an exquisite widow. In old China widows were discouraged from remarriage. Although most of them failed of the ideal, in this case the general rescued the widow from downfall, thereby upholding the age-old tradition and consequently achieving great merit. But the rest of the ten volumes is devoted to events which have nothing to do with what one hesitates even to call romance.

Chinese pornographic novels do not portray romance, as Americans understand the term: the existence of a mutual devotion beyond the sexual plane, a situation where union between a man and woman is the catastasis. Chinese pornography describes the plain externalized pleasures of sex in flagrant detail, and it is interspersed throughout with homosexuality and other forms of perversion. To my knowledge, the only American novel which can faintly compare in frankness is Henry Miller's *Tropic of Cancer*. Even in this book, the author merely uses some artless four-letter words over and over again, relying upon the usual devices such as shadow, sound, implication, or dialogue to portray the sexual act.

All other Western literary pieces, whether banned or not, are true to the pattern of dealing with sex by nuance. James T. Farrell's *Studs Lonigan*, Edmund Wilson's *Memoirs of Hecate County*, William Benton's *This Is My Beloved*, James Jones' *From Here to Eternity*, and Erskine Caldwell's *God's Little Acre* are not to be excepted. The relatively hesitant writers will leave a man and a woman at the chamber door. The bolder ones will step as far as the bed. But the Chinese authors who deal with this matter go relentlessly forward from that point where their Western brethren take leave of the scene.

Finally, from Dr. Hsu's comparative study *Clan, Caste, and Club*,* we have made use of his "tableaux" showing the basic orientations of the cultural systems of China, India, and the United States. These show, in regard to China, the significance of *situational* and of *kinship* factors.

* *Clan, Caste, and Club: A Comparative Study of Chinese, Hindu, and American Ways of Life* (Princeton, N.J.: Van Nostrand, 1963), pp. 163, 172, 205.

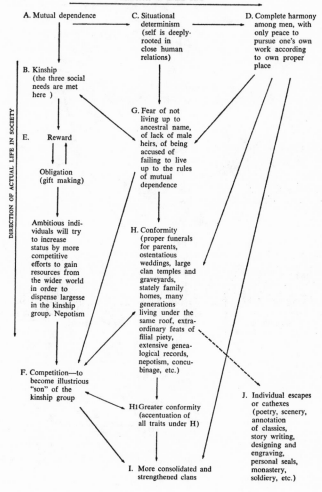

CHINESE ORIENTATION
*Kinship takes precedence over all other ties*

DIRECTION OF IDEAL LIFE

A. Mutual dependence

C. Situational determinism (self is deeply-rooted in close human relations)

D. Complete harmony among men, with only peace to pursue one's own work according to own proper place

B. Kinship (the three social needs are met here )

E. Reward

Obligation (gift making)

Ambitious individuals will try to increase status by more competitive efforts to gain resources from the wider world in order to dispense largesse in the kinship group. Nepotism

G. Fear of not living up to ancestral name, of lack of male heirs, of being accused of failing to live up to the rules of mutual dependence

H. Conformity (proper funerals for parents, ostentatious weddings, large clan temples and graveyards, stately family homes, many generations living under the same roof, extraordinary feats of filial piety, extensive genealogical records, nepotism, concubinage, etc.)

F. Competition—to become illustrious "son" of the kinship group

H1 Greater conformity (accentuation of all traits under H)

I. More consolidated and strengthened clans

J. Individual escapes or cathexes (poetry, scenery, annotation of classics, story writing, designing and engraving, personal seals, monastery, soldiery, etc.)

DIRECTION OF ACTUAL LIFE IN SOCIETY

175

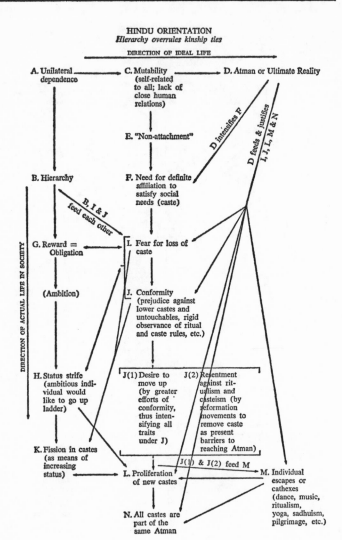

HINDU ORIENTATION
*Hierarchy overrules kinship ties*

DIRECTION OF IDEAL LIFE

A. Unilateral dependence

C. Mutability (self-related to all; lack of close human relations)

D. Atman or Ultimate Reality

E. "Non-attachment"

B. Hierarchy

F. Need for definite affiliation to satisfy social needs (caste)

D intensifies F

D feeds & justifies I, J, L, M & N

B, I & J feed each other

G. Reward = Obligation

I. Fear for loss of caste

DIRECTION OF ACTUAL LIFE IN SOCIETY

(Ambition)

J. Conformity (prejudice against lower castes and untouchables, rigid observance of ritual and caste rules, etc.)

H. Status strife (ambitious individual would like to go up ladder)

J(1) Desire to move up (by greater efforts of conformity, thus intensifying all traits under J)

J(2) Resentment against ritualism and casteism (by reformation movements to remove caste as present barriers to reaching Atman)

K. Fission in castes (as means of increasing status)

J(1) & J(2) feed M

L. Proliferation of new castes

M. Individual escapes or cathexes (dance, music, ritualism, yoga, sadhuism, pilgrimage, etc.)

N. All castes are part of the same Atman

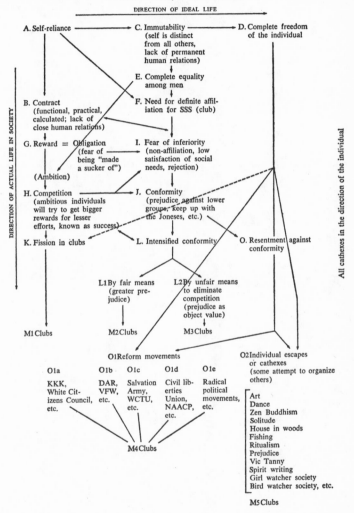

AMERICAN ORIENTATION
*Contract relationship dominates all other*

DIRECTION OF IDEAL LIFE

DIRECTION OF ACTUAL LIFE IN SOCIETY

All cathexes in the direction of the individual

A. Self-reliance

C. Immutability
(self is distinct
from all others,
lack of permanent
human relations)

D. Complete freedom
of the individual

E. Complete equality
among men

B. Contract
(functional, practical,
calculated; lack of
close human relations)

F. Need for definite affiliation for SSS (club)

G. Reward = Obligation
(fear of
being "made
a sucker of")

(Ambition)

I. Fear of inferiority
(non-affiliation, low
satisfaction of social
needs, rejection)

H. Competition
(ambitious individuals
will try to get bigger
rewards for lesser
efforts, known as success)

J. Conformity
(prejudice against lower
groups, keep up with
the Joneses, etc.)

K. Fission in clubs

L. Intensified conformity

O. Resentment against
conformity

L1 By fair means
(greater pre-
judice)

L2 By unfair means
to eliminate
competition
(prejudice as
object value)

M1 Clubs

M2 Clubs

M3 Clubs

O1 Reform movements

O2 Individual escapes
or cathexes
(some attempt to organize
others)

O1a

KKK,
White Cit-
izens Council,
etc.

O1b

DAR,
VFW,
etc.

O1c

Salvation
Army,
WCTU,
etc.

O1d

Civil lib-
erties
Union,
NAACP,
etc.

O1e

Radical
political
movements,
etc.

Art
Dance
Zen Buddhism
Solitude
House in woods
Fishing
Ritualism
Prejudice
Vic Tanny
Spirit writing
Girl watcher society
Bird watcher society, etc.

M4 Clubs

M5 Clubs

177

*Three*

# THE PSYCHOLOGY OF
# JAPAN

Our senior consultant for Japanese psychology is Professor Koji Sato, of Kyoto University, who, in addition to his expertness in Zen Buddhism, is broadly familiar with modern Japanese psychology, both experimental and clinical, and is the editor of the journal *Psychologia.* He has been assisted by a clinical psychologist, Dr. Tadashi Tsushima, who gave us his intensely personal welcome in Kyoto in 1960 and guided our efforts to observe and understand. He has also enlisted the assistance of Professor Y. Hisaki.

## BUDDHISM

The following selection by Professors Hisaki and Tsushima give us some understanding of the psychology of Japanese Buddhism.

### How Japanese Buddhism Has Understood "Mind"

Buddhism, which has played a great part in forming the mentality of the Japanese people, was introduced from China by way of Korea in the sixth century. Thereafter, till the thirteenth century, various Indian and Chinese Buddhist philosophies were successively imported into Japan, mainly by Buddhist priests who had been sent to China, and their principles were gradually absorbed. At first Buddhism was regarded by many as a kind of magic. The government encouraged Buddhism, and it was for the purpose of fostering the happiness of the state and nation, and especially of the governing class, that temples were built and magical rites performed.

Prince Shotoku (574–622), the well-known Buddhist patron, built the Horyuji Temple, the oldest temple in Japan, and exhorted people to believe in Buddhism. In his "Decree of Seventeen Articles," he said: "Believe devoutly in *triratna* [three jewels]: namely, Buddha, Dharma [doctrine], and Sangha [order]. For these are the last resource of all the beings in the universe and the supreme object of faith in all countries." But at the same time, he had a profound understanding of the nature of the human mind. He asserted the relativity of mind, and argued for the forgiveness of sin on the basis of this relativity, saying, "Give up anger. Don't be angry at others' mistakes. Everyone has mind; mind has opinion. If the other's opinion is right, yours is wrong. If yours is right, the other's is wrong. You are not always wise; he is not always stupid. Both are *bombu* [non-sage, mediocre men]. Who can decide which is right? When an issue is important, ask the opinions of the people before deciding." Tolerance is one of the Buddhist virtues, and is based on the awareness of human finiteness which leads to the forgiveness of sin. Among Japanese Buddhists, Prince Shotoku was the first to recognize this.

As mentioned above, Buddhism was mostly accepted as a kind of magic, but eventually morality came to be emphasized, especially by the intellectuals. Kibino Makibi, one of the greatest secular scholars in eighth-century Japan, in identifying *panca silani* (Buddhism's five precepts) with *gojo* (Confucius' five cardinal virtues), stressed the morality of Buddhism. Isonokamino Yakatsugu, Makibi's contemporary, likewise advocated a synthesis of Buddhism and Confucianism. At the beginning of the ninth century, Kyokai wrote *Nipponryoili* (A Collection of Japanese Stories in accordance with the Buddhist Doctrine of Retribution) in order to illustrate the principle: "Cease to do evil; learn to do good." As shallow as it was, this doctrine of retribution was widely accepted by the Japanese people and deeply influenced their ideas of morality.

Buddhist doctrines were studied by priests, and six schools, or sects, were established in the eighth century. One of them, the Hosso, or Yuishiki (Consciousness-Only), school, which was founded in fourth-century India by Vasunbandhu and his brother Asanga, was transmitted to China by Hiuen Tsiang in the seventh century. Though metaphysical rather than scientific, it offered a unique psychological theory.

The six schools of the eighth century, however, had little influence upon the Japanese people. The main streams of Japanese Buddhist history from the ninth century onward were channeled into two newly established sects—Tendai and Shingon, founded, respectively, by Saicho and Kukai, two priests who had been sent to China for study. Saicho (767–822) imported the Chinese *Tien Tai* doctrine, one of whose fundamental ideas is a metaphysical interpretation of the relation between the human mind and the world. It is called "three thousand realms in one mind." The universe is divided into ten worlds. These ten worlds are in one mind. Each world reflects ten worlds, one to another. So one hundred worlds are in one mind. Now, each world has thirty realms; one hundred worlds have three thousand realms. Consequently, these three thousand realms are, after all, in one mind. "Three thousand realms" means all the beings in the universe. Therefore "three thousand realms in one mind" means that the human mind represents and embraces all the beings in the universe.

Accordingly, a profound insight into the human mind led to an insight into the universe and into absolute truth. The *Tien Tai* doctrine required the practice of *shikan* (*samotha* and *vipassana*, spiritual rest and insight) for the attainment of absolute truth. Saicho himself deeply realized the sinfulness of the human mind. In his youth he wrote: "I am the most foolish of all the fools, the most lunatic of all the lunatics. I am sinful enough to be denied salvation." Nevertheless, he became convinced that the possibility of attaining absolute truth lay in sinful mind itself, because, however sinful it might be, the human mind represented the universe.

The all-embracing character of Japanese Buddhism appears in another way in the doctrine of the Shingon sect. Kukai (774–835), who introduced esoteric Buddhism to Japan and was the founder of the Shingon sect, wrote *Jujushinron* (A Treatise on the Ten Stages of Religious Consciousness), which may, in a sense, be compared with Hegel's *Phänomenologie des Geistes*. In the Mahavairocana Sutra he analyzed the logical development of the human mind, and found in the mind ten stages: (1) the mind of a non-sage, or mediocre man, who is like a sheep; (2) the mind of a stupid child acting in terms of morality; (3) the mind of a child realizing "awelessness"; (4) the mind realizing not *anatman* but *skandhabhava* (there is no ego; only the elements of the ego and world exist); (5) the mind eradicating *avidya* (fundamental ignorance); (6) the mind realizing Mahayana Buddhist mercy; (7) the mind realizing *anutpada-avirdha* (non-generation and non-extinction); (8) the mind realizing the a priori way to the absolute truth; (9) the mind realizing ultimate *asvabhava* (non-entity); and (10) the mind comprehending the deepest secret.

1. The first and the second stages represent non-religious consciousness. "A sheep," in Buddhist terminology, means a stupid, greedy man. In ordinary life, we men are subject to various desires, and as we know nothing of the transitory nature of these desires and their ends, we submit ourselves to our desires in vain and pursue the transitory ends.

2. The second stage of the mind, which emerges through the negation of the first one, represents secular morality. Morality is brought into existence, more or less, through negating or controlling our desires. But to do good is not always to be wise, for, from the religious point of view, human goodness is not at all the same as wisdom or attainment of absolute truth. Thus this stage is likened to the mind of a stupid child.

3. In the rejection of secular morality, there comes the third stage,

the consciousness of non-Buddhist religion, especially of Brahmanism. In this stage, morality is endowed with religious significance. Good actions are regarded as the way to a happier future life. Those who are faithful to Deva (God) and do good or practice meditation are believed to be reborn in a happier, aweless state. From the Buddhist point of view, this state, however, cannot be taken for the ultimate end. It is natural that mind which pursues this state should still be compared to that of a child.

4. The fourth and the fifth stages represent Hinayana Buddhism. In non-Buddhist religions, both God and man are deemed substantial, but, in Buddhism, whether in Hinayana or in Mahayana Buddhism, neither God nor man is thought to be an entity. All Buddhists insist that the ego is not an entity but an aggregate of *skandha* (elements). In Hinayana Buddhism, however, this "egoless" (*anatman*) doctrine is mostly accompanied by "the elements" (*skandhabhava*) doctrine. Most Hinayana Buddhists analyzed the elements of the world and the ego, and concluded that these elements themselves were the real entities. Though this is the natural conclusion of their analytical speculation, it tends to obscure *anatman* doctrine. For, when the elements are regarded as entities, it becomes difficult not to recognize the entity of the ego which is essentially the aggregate of them.

5. Therefore some Hinayana Buddhists were concerned about the *pratitya-samutpada* (interdependence of all beings) doctrine rather than the *skandhabhava* doctrine. According to this doctrine, all beings are dependent on one another. But man is ignorant of their independence, and greedily pursues them *as if they were independent entities*. This fundamental ignorance is the basic cause of all the sufferings in our life. Consequently, to eradicate this ignorance or to realize the correlation of all beings must be the end of our life.

6. The eradication of the fundamental ignorance, however, is not the ultimate end of Buddhism, so long as it is pursued self-seekingly. To eradicate the fundamental ignorance self-seekingly is in itself a contradiction, as this eradication is nothing but the destruction of the attachment to everything, especially to self. Buddhism was from the beginning altruistic or philanthropic, and Mahayana Buddhists advocate such altruistic virtues as benevolence, tolerance, and mercy. The sixth stage represents this Mahayana Buddhist mercy, especially that of the Hosso school. According to Hosso doctrine, self is nothing because it is the product of ever-changing *alayavijnana* (states of consciousness), and the elimination of self inevitably leads to altruism.

7. The seventh stage, with its derivatives, also represents Mahayana Buddhism, or its various schools. In Hosso doctrine, self and all beings are negated as they are produced from *alayavijnana*, but the negative

function of mind itself is not yet negated. On the contrary, in the seventh stage which represents Sanron school doctrine, this negative function is also negated, and mind is regarded as *sunyata* (void). *Sunyata* is not nothingness, but no-thing, that is, neither generative nor extinctive. In this sense, it is called "the Middle Way." Strictly speaking, *sunyata* means the absence of any predicate, and can be expressed only through the way of negation, *via negationis*.

8. But on the *via negationis* such negativeness itself must be finally negated. The ultimate goal of the *via negationis* is the negation of negation, or affirmation. Thus it is realized in the eighth stage that void is existence as well as that existence is void. This represents the teaching of the Tendai sect. According to Tendai doctrine, this realization is the way to the absolute reality (absolute truth) and this way is not found either by Buddha or by man. It is a priori in the sense of being inaccessible to all human approaches.

9. Nevertheless, this realization is only an introduction to the absolute reality, considered from the higher standpoint, the ninth stage, which represents the doctrine of the Kegon school. Kegon doctrine teaches that the absolute reality unfolds itself into all the beings in the universe in numerous ways, and all the beings depend upon the interrelationship and constitute a harmonious cosmos. Therefore, neither these interdependent beings nor the absolute reality, the whole body of them, is an entity. By realizing this ultimate nonentity, it becomes possible to attain absolute truth, or to hold communion with the harmonious cosmos.

10. But the possibility of this communion is not yet actualized in the ninth stage. Only in the tenth stage, which represents Shingon doctrine, can this communion be achieved. According to Shingon doctrine, the absolute truth is personified as Buddha Dainichi (Mahavairocana: The Great Illuminator), and all the phenomena in the universe are regarded as activities of his body, speech, and mind. These three aspects of Dainichi are called "three secrets," because they cannot be known directly. Therefore, communion with Dainichi is held only by practicing certain kinds of sacraments which represent various functions of these "secrets." Thus man can make himself identical with Buddha, or become Buddha. This is the final goal of the whole of Buddhism.

First of all, these ten stages, or steps, show the logical development of the human mind; by traversing them, one by one, absolute truth is attained. Each stage has its limit, and when it is reached, that stage is negated and the spirit can climb to the next stage. Thus, each stage conveys a sense of the negative value of the preceding one. In other words, every stage implies a contradiction; and it is through the realization of this contradiction that the higher stage reveals itself. So the development of mind through the ten stages may be said to be dialectical.

Secondly, from the standpoint of Shingon doctrine, the first nine stages are considered to be embraced by the tenth stage. As Buddha Dainichi embraces all the phenomena in the universe, so these stages are also embraced by him. Each stage, though distinguished from all the others, is, at the same time, an expression of Buddha himself. Each stage is the self-determined form of Buddha, the absolute reality. Even the greedy, stupid mind—the mind of "a sheep"—is, in essence, mediated by the absolute reality. Here we find one of the most thoroughly pantheistic of all Buddhist doctrines.

Thirdly, the stages illustrate rather well the historical development of Buddhism. The last seven stages, as indicated in the outline above, represent the doctrines of various Buddhist schools. And as the sequence of these stages almost agrees with that of the origination of those schools, they tell the history of Buddhist thought.

Finally, Kukai's treatise on the ten stages of mind is an explication of Shingon apologetics as well as the history of Buddhist thought. He gave the Shingon doctrine the highest position among all the Buddhist philosophies, and in this way attempted to establish the superiority of Shingon doctrine.

Nevertheless, in spite of Kukai's efforts, Shingon gradually lost its philosophical superiority and underwent a degradation, even though it appealed to the Japanese people and spread rapidly among them. Shingon is essentially magical and esoteric. Its doctrine teaches that the magical rites lead not only to communion with Buddha but to the various worldly goods—material possessions, recovery from illness, bountiful harvests, plentiful rain, and so forth—and for this reason Shingon spread widely. The magical rites, however, were adopted by the Tendai sect and had some effect upon Zen and Jodo Buddhism as well.

Jodo (Pure Land) Buddhism was introduced to Japan in the sixth century, but it did not take hold until about the end of the tenth century. As the ancient dynasty declined, there spread a kind of eschatology, a belief that the end of the world was drawing near. People earnestly sought salvation through Jodo Buddhism, which taught that man could be delivered from sin through faith in Buddha Amida (Amitabha: The Lord of the Infinite Light and Life) and the practice of *nembutsu* (contemplation of Amida or recitation of his name). Jodo Buddhism enjoyed its greatest flowering during the lifetime of Honen (1132–1212), the founder of the sect.

According to the Jodo Buddhist myth in the Skhavativyuha Sutra, Amida made a vow that he would help everyone who asked him for salvation and practiced *nembutsu*. On this original vow Honen laid special stress in his *Senchakuhongannembutsushu* (Extracts and Explanations of the Sutras and Treatises concerning the Original Vow of Amida). In this book he developed his unique thoughts and laid the philosophical foundation for Jodo Buddhism: (1) any non-sage, or mediocre man, can share in Amida's salvation, (2) through the practice of *nembutsu*, (3) if he possesses *sanjin* (three minds, three aspects of faith), (4) for *nembutsu* is in accordance with the original vow of Amida.

1. Honen emphasized the universality of Amida's salvation and preached his mercy. Since salvation by self-effort, as Tendai or Shingon doctrine taught, was beyond the powers of non-sage, or mediocre, men, he advocated salvation through another power, Amida. It was for this reason that many of the helpless, common people came to believe in Jodo Buddhism.

2. The means of salvation must be easy for such people. He limited *nembutsu* to the recitation of Amida's name, rejecting the contemplation of Amida. In any case, he regarded contemplation merely as an aid.

3. Recitation, however, is not merely the utterance of an incantation. To recite Amida's name, or to chant *Namu Amidabutsu* (adoration to Buddha Amida) is an expression of faith in him. Honen asserted that salvation is attained through faith alone, and insisted upon the importance of the three aspects of faith, which were set forth in the Amitayurdhyana Sutra—namely, *shijoshin* (sincerity), *jinshin* (devoutness), and *ekohotsuganshin* (religious decision).

4. But one may inquire how salvation can be attained through faith in Amida and recitation of his name. Honen's answer is as follows: *nembutsu* is in accordance with the original vow of Amida, and Amida selected *nembutsu* as the only means to salvation, because the merit of *nembutsu* is absolute and it is easiest to practice; Amida desires the welfare of all human beings, as he is infinite benevolence itself.

After Honen's death, Jodo Buddhism was splintered into several sects. Shoku (1177–1247), one of the ablest disciples of Honen, developed Honen's doctrine and founded the Seizan Jodo sect. He emphasized comprehension of the three aspects of faith, through which, he taught, man could share in Amida's salvation. In his letter to the Empress Dowager, he interpreted *sanjin* in a unique way: sincerity is not man's sincerity but his seeking for Amida's sincerity, from which *jinshin* and *ekohotsuganshin* arise. Shan Tao, a famous Jodo Buddhist in seventh-century China, had already divided *jinshin* into two aspects: awareness of sin and faith in the mercy of Amida. Shoku pointed out the interrelation of these two aspects of *jinshin*—the more deeply one realized his own sin, the more profoundly he realized the mercy of Amida, and the converse was also true. In this way he asserted that faith in Amida is the only path to salvation, and he regarded self-effort itself as useless for salvation. At the same time, according to Shoku, once a man has faith in Amida, his acts are endowed with religious significance, though they are in themselves useless and insignificant. This is not, however, the affirmation of self-effort; it is Amida's power which gives significance to the insignificant.

In a somewhat different way, Shinran, one of the disciples of Honen, and the founder of the Jodo-Shin sect, also emphasized the power of Amida and the powerlessness of human beings. He deeply realized the sinfulness of men. In his *Kyogyoshinsho* (Selections and Commentaries on the Sutras and Treatises concerning Jodo Buddhist Teaching, Practice, Faith, and Attainment), he confessed his own sinfulness: "Alas! How deplorable it is that I should not be glad to share in Amida's salvation, nor pleased with drawing near the true attainment, drowning in the great oceans of passions and wandering about the high mountains of fame and profit! I am ashamed of this, and exceedingly grieved."

He carried the idea of sin to the extreme conclusion: In *Tannisho* (Analects of Shinran), he writes: "Even a good man will be helped, how much more a bad man!" This paradoxical saying, however, does not mean an affirmation of evil. "A good man" is not one who does good, but one who has little self-consciousness of sin; whereas "a bad man" is one who realizes his own sinfulness. This points up three facts: first, that goodness and badness are conceived in terms of religion, not of morality, and that the ultimate badness or sin is *boho* (calumny of Buddha's teachings), which arises from self-will through which man rebels against Amida; second, that man's acts are in themselves sinful, even though they may be regarded as good from the standpoint of morality; third, that all human beings are sinners, whether they realize it or not, and whether they are virtuous or not. In short, sin, meaning religious alienation from Buddha, is native to man. This is, as it were, a doctrine of original sin, and morality has thus lost religious significance altogether.

Though Shinran realized the sinfulness of human beings most profoundly, his idea of sin was not always properly understood. It has often been interpreted as the negation of morality, and his doctrine was considered to be ethically weak. Rennyo (1415–1499), Shinran's descendant in the eighth generation, was said to have compensated for this weakness by modifying Shinran's doctrine; Rennyo added to Shinran's teachings the commandment that men should practice the worldly moralities and be loyal to their feudal lords and to the patriarchs of their families. Rennyo's addition did not reinforce Shinran's doctrine, but threw it into confusion, for Shinran's doctrine of original sin was contradictory to Rennyo's commandment.

According to Shinran, we have seen, we men are sinful and our practice of either morality or contemplation is not the way to salvation, which can be opened by faith alone. We have nothing, whereas Amida has all. Even the power to put faith in Amida is not native to man, but a gift from Amida. He is a Faith-Giver as well as Savior. This concept of faith as a gift of Amida is peculiar to Shinran's doctrine which emphasizes self-awareness of sin. No Buddhist before him had had this idea, which was the first consequence of his doctrine of original sin. The second consequence of his doctrine was his distinction between his position and that of priests who advocated self-effort and were little aware of their sinfulness, and that of laymen who lacked the awareness of sin altogether. He called his own position *gutoku*, which means "the fool" or "the non-priest and non-layman." It was on this basis that he married openly and above all rejected hypocrisy. Shan-Tao had taught: "Don't allow deceitfulness in your mind while behaving outwardly like a wise, good, and diligent man!" Shinran turned this about, to read: "Don't behave outwardly like a wise, good, and diligent man, because you really have deceitfulness in your mind!" Indeed, Shinran can be said to have comprehended the abuse of the human mind.

Shinran was the first of the Japanese Buddhists to reject magical rites. As we have already noted, Buddhism was regarded as a kind of magic when it was first introduced to Japan. Later the magical rites of Shingon were taken up by the other sects. Even Judo Buddhism could not rid itself of their influence. Honen often performed *jukai* (the Buddhist initiation ceremony) for noblemen, as it was considered to have magical efficacy in helping one recover from illness. Shoku did not reject the performance of magical rites. Great as the influences of Shingon upon Jodo Buddhism were, Shinran isolated himself from them, preaching that faith in Amida and the recitation of his name were the only assurance of blessedness.

In addition, his doctrine is monotheistic, or at least henotheistic. Buddhism, of course, was originally pantheistic, and on this basis developed a polytheistic trend. Most Buddhists worshipped numerous Buddhas, Bodhisatvas (Buddhas-to-be), deities, and demons. Moreover, at the beginning of the tenth century, the deities of Shintoism began to be identified with those of the Buddhist pantheon. For example, Amaterasu (Great Sun Goddess) was identified with Buddha Dainichi (Mahavairocana); Futsunushi (a war god) with Buddha Yakushi (Bhaisajyaguru); Amenokoyane (the ancestral god of Fujihara clan) with Bodhisatva Jizo (Ksitigarba); and Hachiman (a war god) with Buddha Amida. This syncretic trend contributed greatly to the popularization of Buddhism. But Shinran was antagonistic to the syncretic or polytheistic trend, teaching faith in Amida alone, though he did not necessarily deny the existence of numerous deities. In this way he tried to assure the purity of faith in Amida.

The syncretic trend, however, had a considerable effect upon the Jodo Buddhists. One of them—Ippen (1239–1289), a disciple of Shoku and the founder of the Ji sect—believed that he had been given faith in Amida by Izanagi (a god of creation), the deity of the Kumano shrine, and that Amida had incarnated himself as Izanagi in order to popularize Jodo Buddhism. It goes without saying that his belief is a result of syncretism. He also danced in ecstasy at the recitation of Amida's name; this ecstatic dance is nothing but a remnant of old shamanistic Shintoism.

Though Ippen's conciliatory attitude toward the popular belief made a great contribution to the spread of his teachings and his adherents were said to number 250,000, other characteristics can be pointed out in his doctrine. One of them is a mystical style of thought which resembles St. Paul's. In a letter to one of his followers, Ippen wrote: "The moment I put faith in Amida, realizing how brief our life is, I am no longer what I used to be. My heart is identical with Amida's heart, my behavior with Amida's, and my life with Amida's." This is a further development of Shoku's idea that the adoration of Amida means entrusting him with our own lives. Shoku's "spiritual passivity" was extended and transformed into spiritual activity by Ippen. This mystical identification of Amida and his devotees is intimately connected with Ippen's pantheistic doctrine. "When I repeat *nembutsu* over and over again," he wrote in another letter, "there is neither Buddha, nor myself, much less various reasonings. . . . All the creatures and all the beings in the world, even the blowing wind and roaring waves are nothing but *nembutsu*." *Nembutsu*, as it were, embraces the universe, and all the phenomena and existences in the universe are manifestations of *nembutsu*. True faith is, in reality, nothing but incorporation into this all-embracing *nembutsu* itself. His syncretic tendency is based on this pantheistic doctrine, and Shinran's henotheistism has faded out.

As mentioned above, Jodo Buddhism embraced various trends, but it had important effects upon the life and culture of the Japanese people. "Jodo" originally meant the paradise after death, and Amida's salvation has usually been taken for rebirth to Jodo. But Shoku and Shinran did not necessarily regard salvation as rebirth to Jodo. Though not reflecting the traditional meaning of Jodo, they asserted that man could share in Amida's salvation in this life. But their idea of salvation was not widely accepted. For the common people, Jodo was, to the end, paradise after death, and they believed that Amida had promised this paradise to his devotees. This belief not only contributed to the spread of the idea of *anitya* (mutability), that life is temporary, but also provided the people with the consolation that a happier life was promised after death, no matter how miserable this life might be.

Such a consolation, however, is only a form of resignation to this life, and the effort to overcome difficulties and to improve the conditions of life is thus given up. But sometimes the conviction of rebirth to Jodo exerted another influence. It swept away fear of death, and it was both this conviction and the equalitarian doctrine of the Jodo-Shin sect that formed the ideological basis of the revolt, in the fifteenth century, by the peasants who belonged to this sect. Jodo Buddhism influenced other sects of Buddhism, too: Amida-worship flowed into the Tendai and Shingon sects. Even *shodai* (the recitation of *Namu Myohorengekyo*: adoration to the Saddharma-Pundarika Sutra) in the Nichiren sect, founded by Nichiren (1222–1282), was an imitation of *nembutsu*, though Nichiren himself criticized and furiously opposed Jodo Buddhism.

As greatly as Jodo Buddhism influenced the various aspects of Japanese life, and as profoundly as it comprehended the human mind, it ceased to advance its ideas after the fourteenth century. The same is true of other Buddhist sects. Especially after Buddhism had gained the position of a state religion in the seventeenth century, the feudal hierarchy smothered its religious spirit. After the Meiji Restoration (1868), State Shintoism occupied the supreme position among all the religions in Japan. Buddhism lost government patronage, and had to face a new, powerful rival: Christianity.

There then appeared two trends in Buddhism: one, the movement to regain government patronage by cooperating with the government's policies and even with State Shintoism; the other, the effort to advance and modernize the doctrines of Buddhism by absorbing new, Western ideas. The former represented the mainstream of Buddhism in Japan after the Meiji Restoration, whereas the latter was a sporadic development. From the philosophical point of view, however, the trend toward modernization was far more important. One of its representatives was Manshi Kiyosawa (1863–1903), a priest of the Jodo-Shin sect. Kiyosawa threw light upon Shinran's doctrine by confronting it with European philosophies, especially those of Hegel and Epictetus, in which he was erudite. He called his own philosophy *seishinshugi*, which means spiritualism. Its objectives were: (1) to seek spiritual satisfaction, (2) to seek the satisfaction of others as well as one's own, and (3) to seek spiritual freedom which does not contradict dependence.

(1) According to Kiyosawa, Amida is not an objective, empirical being, but a purely spiritual reality which embraces all the existences in the universe. Therefore the only way to put faith in Amida is by introspection, and through this introspection one can attain spiritual

satisfaction. (2) This satisfaction, however, is accompanied by co-satisfaction with others, because faith in Amida requires abandonment of egotism. (3) On the one hand, faith in Amida means obedience to him or dependence upon him. On the other hand, one who puts faith in him is free from spiritual finiteness, because through this faith one can be united with Amida, the infinite. Here dependence, paradoxically, leads to freedom.

Kiyosawa thus regarded society as abominable, emphasizing only spiritual satisfaction through realizing introspectively the sinfulness of human beings and putting faith in Amida. But the lack of social interest is one of the defects in Kiyosawa's ideas and, at the same time, of Japanese Buddhists proper, though they have looked deeply into the psychology of the individual and have comprehended the nature of the human mind. This is one of the problems that Japanese Buddhism must solve in the future.

## ZEN BUDDHISM

There had always been an "inner" Hinduism and a popular or practical Hinduism. So, too, as we have seen, Buddhism had always presented a practical, or action, side and a spiritual, or subjective, side. In China, over a thousand years ago, the issue led to the growth of Zen, a school seeking to find in the *self*, rather than in ceremony or doctrine, the path to wisdom.

In the following selection,* Paul Reps, introducing the American reader to Zen, seizes the inner message: the less *done*, the clearer is reality; the less *said*, the more direct our apprehension. Bodhidharma's pupils approach nearer and nearer to understanding as they have less and less to say.

The first Zen patriarch, Bodhidharma, brought Zen to China from India in the sixth century. According to his biography recorded in the year 1004 by the Chinese teacher Dogen, after nine years in China Bodhidharma wished to go home and gathered his disciples about him to test their apperception.

Dofuku said: "In my opinion, truth is beyond affirmation or negation, for this is the way it moves."

Bodhidharma replied: "You have my skin."

* From Paul Reps (compiler), *Zen Flesh, Zen Bones* (Garden City, N.Y.: Doubleday & Co., 1961), pp. xiv–xv.

The nun Soji said: "In my view, it is like Ananda's sight of the Buddha-land—seen once and forever."

Bodhidharma answered: "You have my flesh."

Doiku said: "The four elements of light, airiness, fluidity, and solidity are empty [i.e., inclusive] and the five *skandhas* are no-things. In my opinion, no-thing [i.e., spirit] is reality."

Bodhidharma commented: "You have my bones."

Finally, Eka bowed before the master—and remained silent.

Bodhidharma said: "You have my marrow."

Old Zen was so fresh it became treasured and remembered. Here are the fragments of its skin, flesh, bones, but not its marrow—never found in words.

The directness of Zen has led many to believe it stemmed from sources before the time of Buddha, 500 B.C. . . . The problem of our mind, relating conscious to preconscious awareness, takes us deep into everyday living. Dare we open our doors to the source of our being? What are flesh and bones for?

We now turn to our consultant Koji Sato, professor at the Imperial University of Kyoto, to introduce us to the beauty and present meaning of Zen Buddhism.

### How Zen Conceives of Mind

Master Eisai (1141–1215), founder of the Japanese Rinzai sect of Zen Buddhism, praised the mind in his preface to *Defending the Nation by Promoting Zen*:

"Great is the Mind. Heaven is so high that it cannot be reached, and the Mind goes above Heaven. The light of the Sun and the Moon is not to be surpassed, and the Mind goes beyond the light of the Sun and the Moon. Thousands of thousands of worlds are not to be covered, and the Mind goes beyond the thousands of thousands of worlds. There is the Great Emptiness, there is the Fundamental Spirit. The Mind includes the Great Emptiness in it, and conceives the Fundamental Spirit. Through Me (the Mind), Heaven covers the world and the Earth carries it. Through Me, the Sun and the Moon go round; through Me, the four seasons change; through Me, all creatures grow. Great is the Mind. We dare to name it, even though originally undefinable, and call it 'The Highest Vehicle,' 'The First Principle,' 'The Reality of Prajna,' 'One True World of Dharma,' 'The Highest Bodhi (Enlightenment),' 'Ryogon Samadhi,' 'The Eye and the Treasury of the Right Law,' 'The Mystery Mind of Nirvana.' "

Thus conceived, the Mind is the highest function which can embrace the whole universe, including beings and non-beings, from man even to God. It may be compared with the glory of the *pensée* of Pascal. Shinichi Hisamatsu, a modern Zen philosopher (1889–    ), in discussing the Oriental Nothingness, mentioned as one of its characteristics the "Mind-in-Itself" nature of Oriental Nothingness. According to his view, "such phrases in Zen as 'Right-Dharma-Eye-Treasure-Nirvana-Wondrous-Mind,' 'directly pointing to the Mind of man, realizing its Nature and attaining Buddhahood,' 'transmission from Mind to Mind,' 'Pure Mind,' 'Unattainable Mind,' 'Mind-Dharma,' 'Mind-Nature,' 'Mind-Source,' 'Mind-Ground,' and 'Mind-Itself' all express this 'Mind-in-Itself' nature of Oriental Nothingness. Not only in Zen, but in Buddhism in general, though the True Buddha is likened to empty space, it is nevertheless said that Buddha is 'Mind-Itself' of 'Awareness-Itself,' attesting to the fact that Buddha is mind-like.

"Although Oriental Nothingness is said to be mind-like, it is not what we ordinarily call mind. When saying that Oriental Nothingness is mind, Mind, as we have said, is likened to empty space. For this Mind is Mind possessing all the characteristics of empty space: unobstructedness, omnipresence, impartiality, broadness and greatness, formlessness, purity, stability, the voiding of being, the voiding of voidness, unattainability, 'one-aloneness,' having neither an interior nor an exterior, and so on. Since what we ordinarily call mind does not possess these characteristics of empty space, in order to distinguish clearly between the two, it has, from ancient times, been said that this 'Mind is like empty space.'

"When it is also said that the True Buddha, like trees and rocks, is 'without mind,' 'without thought,' 'free from thought,' 'not caught up in the thinking of good and evil,' 'not-thinking-itself' of 'the stopping of the functioning of mind, thought, and consciousness,' and when it is said that true knowing is 'being free from knowing,' or that true awareness is unawareness, or that True Nature is ungenerated and unperishing, without birth and death—it is not meant that the True Buddha, true knowing, true awareness, and True Nature are all merely like trees and rocks, without mind, self-consciousness, or life. They mean rather that the ordinary mind, self-awareness, and the life which we have are not the true mind, true self-awareness, and true life, and that the true mind, true self-awareness, and true life must possess the characteristics of empty space.

"The True Buddha is not without mind, but possesses Mind which is 'without mind and without thought.' It is not without self-awareness, but possesses Awareness which is 'without awareness'—an egoless ego.

It is not without life, but possesses Life which is ungenerated and un-perishing.

"The ordinary mind has obstructions, places where it does not reach, differentiation, limitation, form, defilement, arising and decaying, dimension, attachments, acquisitions, an interior and an exterior, and is un-collected. One generally has such a mind as subject, and therefore one is an ordinary being and not Buddha. When one composes this mind and returns to the Original-True-Mind, which is like empty space, then for the first time one is oneself Buddha. 'Zen sitting,' in which mind and body have 'fallen off,' is nothing but a state of the realization of such a 'Mind like empty space.' The mind referred to in the Sukhavati-vyuha Sutra is thus described:

" 'The mind is the Buddha; it is Mind which becomes Buddha. You must know that Buddha is nothing but Mind. Outside of Mind there is no other Buddha.' This is precisely 'Mind like empty space.' In speaking of the 'Mind-Itself' nature of Oriental Nothingness, my intention is to indicate that Oriental Nothingness as mind is this 'Mind like empty space.' "

From the Zen point of view, the Fundamental Mind is the True Self and the Buddha. The psychological problem is how to realize this Fundamental Self in ourselves. This is one of the goals of Zen meditation. Master Tetsugen (1630–1682), of the Obaku Zen sect, taught:

"When one practices *zazen* [Zen sitting] and applies the most sincere faith, there arise three kinds of ideas: good, bad, and neutral. The good represents the mind that thinks of good; the bad, the appearance of the bad; the neutral, not good and not bad, but rather absentminded. . . . If one concentrates one's efforts on *zazen* wholeheartedly, without ac-cepting such ideas, and always reinforcing one's intention, the mind ripens a little, and sometimes, for a short while, a state of mind occurs that is occupied with neither good ideas nor bad ideas nor neutral ideas, but rather is as clean and clear as a polished mirror or perfectly clear water. When this happens, one should practice *zazen* more intensively. If one concentrates his efforts on *zazen* diligently, the mind becomes clear for a short time, but gradually the period of time lengthens; and sometimes the mind remains clear for one-third of the time of *zazen*, sometimes for two-thirds of the time. Or the mind may continue to be clear and clean from the beginning to the end—without good ideas or bad ideas, nor neutrally absentminded—like a fine autumn sky, like a polished mirror, like the empty sky, as if the world of dharma lay in the breast and the breast felt incomparably cool. At this point *zazen* is more than half accomplished. In Zen this stage is described in various ways: 'to become one piece,' 'one-colored field,' 'the man at the bottom of the

195

Great Death,' or 'the situation of *fugen*.' But this is not the final stage. 'The place where the mind seems like a clear sky without ideas is the Eighth Consciousness, which is the fundament of all delusions.' 'This consciousness resembles essentially the original mind, but, still dormant in *avidya*, it cannot be called the original mind. On the other hand, it is not mere delusion, for all the delusion ideas have already been extinguished here, even though it cannot be called the proper original mind. If the student reaches this state, he should practice more assiduously, for it is the preparatory stage for the appearance of the true *satori*.' 'Realizing that is it not the final stage, even though the darkness of delusive ideas is gone, if one ceaselessly concentrates on his practice, without exultation, without expectation of *satori*, without special ideas, without special mind (no-mind), the true *satori* presents itself all at once, and everything is illuminated as if a hundred or a thousand suns suddenly appeared.' This is called 'seeing one's own nature and becoming Buddha,' 'having great *satori* from the bottom,' 'nirvana and bliss.' At this time one encounters all the Buddhas of the past, the present, and the future at once, realizes the essence of Sakyamuni and Bodhidharma, witnesses the nature of all beings, digs to the bottom of all creatures and of heaven and earth. The joy is indescribable. 'When one has attained such an enlightenment, everything in the universe is nothing but our original mind.' "

Master Takuan (1573–1645) compared original mind with water and the delusive mind with ice. One cannot wash hands and feet with ice, whereas water can wash everything. "The original mind is the mind which does not stop to abide at one point and pervades the whole body. The delusive mind is the mind which is congealed to one place, fixed on one idea." He also discussed "mindful mind" and "mindless mind" or "no-mind." "Mindful mind is the same as the delusive mind. The mindful mind has something to stick to. The mind of no-mind is the same as the original mind, and it is the mind without fixation, without discrimination and consideration, the mind pervading the entire body, filling it. It is the mind not located anywhere." In a letter to Yagiu Tajima-no-kami, a famous swordsman, Takuan spoke of "where to locate the mind": "Where is the mind to be directed? When it is directed to the movements of the opponent, it is taken up by them. When it is directed to his sword, it is taken up by the sword. When it is directed to striking down the opponent, it is taken up by the idea of striking. When it is directed to defending yourself, it is taken up by the idea of defense. When it is directed to the pose which the opponent assumes, it is taken up by it. At all events, they say they do not know just where the mind is to be directed." Thus "the thing is not to try to localize the mind anywhere but to let it fill up the whole body, let it flow throughout the totality of your being. When this happens, you use the hands when they are needed, you use the legs or the eyes when they are needed, and no time or extra energy will be wasted."

The mind may be psychologically bound. For instance, "one may deliberate when an immediate action is imperative, as in the case of swordsmanship. The deliberation surely interferes and 'stops' the course of the flowing mind. Have no deliberation, no discrimination. Instead of localizing, or keeping in captivity, or freezing the mind, let it go all by itself freely and unhindered and uninhibited. It is only when this is done that the mind is ready to move as it is needed all over the body, with no 'stoppage' anywhere."

When Dogen (1200–1253), founder of the Soto sect of Japanese Zen, returned home from China, where he had studied Zen, he was asked what he brought back with him. It is said that he replied: *"Nyunan-shin"* ("flexible mind" or "soft mind"). This flexible mind is the same as the original mind, which has been compared with water.

Chuho Myohon (1263–1323), a Chinese Zen master, once said: "What is Zen? It is the name of our Mind. What is the mind? It is the body of Zen." This is an extremely abbreviated expression of the relationship of Zen and the mind. This saying may be explained as: "Zen is the way of actualizing the Fundamental or Cosmic Mind."

Hisamatsu adds that "Oriental Nothingness can be said to be Mind" (in the sense previously described). Although we say mind, however, it is not a mind that can be viewed objectively outside of ourselves. It must be inside of ourselves as the subject of ourselves. That is, "it must be such that Mind is Myself and that I am that Mind."

This Mind is our True Self. According to Master Dogen: "To learn Buddhism is to learn the Self. To learn the Self is to forget the Self. To forget the Self is to be authenticated by the universe. To be authenticated by the universe is to make one's Self and others' Self fall off." "Simply letting go, forgetting our body and mind and throwing ourself into the Buddha's house, being conducted from the side of Buddha and behaving accordingly, then, not exerting any force, not expending the mind, we leave behind life and death, and become Buddha."

This means that to study Buddhism is to learn what the True Self is. The True Self can be realized when one becomes one with the Self and when the consciousness of the Self—concretely speaking, the consciousness of the mind and the body—falls off. When the consciousness of the Self falls off, one has reached the highest state of consciousness and everything presents itself as it is, and everything goes by its own law (dharma).

In this highest state of consciousness, there is no subject-object duality. Everything can be identified with the Self. This identification may be called, to use Dewey's terminology, transactional.

The following fragments from Dogen and Takuan provide rich, concrete images with which to flesh out these abstractions.

Dogen: To learn and to authenticate everything by the measure of the self is delusion. When everything comes to contribute to and authenticate the self, it is enlightenment. . . . To learn Buddhism is to learn the self. To learn the self is to forget the self. To forget the self is to be authenticated by all. To be authenticated by all is to make one's self and others' self fall off. . . . Acquiring enlightenment is like the moon reflecting on water: the moon does not become wet, nor is the water ruffled. Even though the moon gives immense and far-reaching light, it is reflected in a pool of water. The full moon and the entire sky are reflected even in a dewdrop in the grass.

Enlightenment does not destroy man, just as the moon does not destroy the water. Man does not obstruct enlightenment, just as a dewdrop does not obstruct the moon in the sky. The brighter the moonlight appears on the water, the higher hangs the moon itself. You should realize that the length of time that the moon appears on the water will testify to the dimensions of the water and the fullness of the moon.

Takuan: The original mind is the mind which does not stop to abide at one point and which pervades the whole body. The delusive mind is the mind which is congealed to one place, fixed on one idea. The original mind becomes the delusive mind when it is congealed and gathered at one point. It is the principle of the original mind to strive not to lose itself, for the original mind, when it loses genuineness, fails to work effectively. Figuratively speaking, the original mind is like water and does not abide at one place, whereas the delusive mind is like ice and the hands and head cannot be washed with ice. Hands and legs can be washed only when the ice is melted into water and can be made to flow. When the mind is congealed to one point and abides on one matter, it is like the ice which remains solid and cannot be used freely. It is as if hands and legs cannot be washed with ice. To use the melted mind like water which spreads over the whole body freely where it goes: this is the way of the original mind.

Mindful mind is the same as the delusive mind. The mindful mind has something to brood upon and to stick to. The mind has something to think about and shows consideration and discrimination; then it is called the mindful mind. The mind of no-mind is the same as the original mind; it is the mind without fixation, without discrimination and consideration, the mind pervading the whole body, filling it. It is the mind not located anywhere. . . . It is not like stone or wood. The mind without abiding is called no-mind. . . . When the mind can achieve no-mind, it does not stop at one matter, does not lack anything, can always work wherever it is needed, like water that is stored. The mind fixed and abiding at one place cannot work freely. The wheel can rotate because it is not fixed to the axle. If it is fixed to the axle it cannot rotate. The mind also cannot work if it is fixed to one point. If one has something in mind, he cannot hear even if he hears. For the mind abides with an idea.

Hakuin (1689–1768) is representative of the Zen Buddhism of recent centuries:

If those who participate in Zen training can enter the state of Great Doubt, it is certain that all of them, one hundred of one hundred, one thousand of one thousand, can break through. When one enters the Great Doubt, it is vast and empty on all sides; it differs from life, differs from death; as if one sits among thousands of miles of layers of ice or in the immense castle of crystal; one feels extremely cool and clean. One forgets to stand up when sitting, forgets to sit down when standing up. There are no ideas or sentiments; there is only *Mu* (Nothingness). It is as if one stands in the vast empty sky. When this happens, if one pushes forward in a breath without fear and reflection, and does not withdraw, one might feel as though the layers of ice had collapsed or the crystal castle had fallen down, and might experience the Great Joy which he had not yet seen or heard in all his life. Life and death, or nirvana, all seem as if yesterday's dream, thousands of worlds seem as if foam of the ocean; the glory of all the sages and saints seem like a flash of lightning. This is called the moment of the Great Enlightenment, Everything Falling Off. It cannot be communicated, cannot be explained; one can only experience it by oneself, as one feels cold or hot. All dimensions are dissolved; the past, present, and future are seen in one ideation. What other Joys are there on the earth and heaven beyond this? One can achieve it without fail in only three to five days if one pushes on with all his might. . . . Abbot Eshin, who was so excellent in his wisdom, virtue, and faith, would have been able to reach Enlightenment in one to two months, at most in six months to a year, if he had studied through *Mu* or other Zen *koans*. Actually he needed forty years of studiousness to achieve Enlightenment, for he practiced only calling *nembutsu* and chanting sutras.

The next selection gives us a recent firsthand account of Zen experience.*

### A Zen Training Session

More than twenty people attended the August training session of the Daikoji Temple, and the training hall was full. I wondered whether the Master would be able to give careful individual guidance. However, I made up my mind to follow obediently the direction of the Master for five days without doubting, relying completely on the Master.

The first, second, and third stages passed quite naturally, no troubling ideas appeared during the sitting from morning to night, and I was at ease in the hope that something would happen. During that time noises were not very annoying, but sometimes my neighbor troubled me by his humming voice, and my warming up seemed to have troubled my neighbor. I had confidence, however, in the success of these first three days at least.

I proceeded to the fourth and fifth stages, and, on the fourth day, some anxiety or irritation arose along with a doubt that, with so little time left, I would experience a change from my previous state of mind even if I continued the training.

During the training, the voices of the other participants began to sound queer to me. But this feeling disappeared soon; only my anxiety remained. In the afternoon of the fourth day I told the Master about it. He encouraged me, saying: "You should not expect something special. Go straight forward!" I renewed my efforts and in the meantime something dark and dusty in my brain seemed to go away with the calling voices. I told that to the Master and he approved it: "It's all right! Straight forward!" I continued my *zazen* practice with all my might; then a feeling such as "that of emptiness and serenity" began to appear. I reported it to the Master and he encouraged me, saying: "Good, good! Straight forward!" And the fourth day came to an end.

In the morning of the fifth day, I got up at five and began to sit. I returned to the state of the previous night. And unexpectedly soon a conversion came. In less than ten minutes I reached a wonderful state of mind. It was quite different from any which I had experienced in *seiza*

* From Shuichi Kitahara, "A Zen Training Session," *Psychologia*, VI, No. 4 (December, 1963), 188–189.

(sitting quietly) or other practices. It was a state of mind incomparably quiet, clear and serene, without any obstruction. I gazed at it. Entering this state of mind, I was filled with a feeling of appreciation, beyond usual joy, on my reaching such state of mind, and tears began to flow from my closed eyes. I had read reports of tears on such occasions, and wondered how sentimental they were. But now I myself could not stop the tears. Warm tears gushed from my eyes. Of course I was choked with emotion. Now and then I wiped my tears with a handkerchief and sat silently until the *dokusan*, or personal interview.

Even more surprising to me was the change that went on in my mind. The calling voices of the participants which had sounded strange and the noises of the cars passing by the Temple which had been annoying, became suddenly quiet. The sound was heard clearly but was no longer annoying. All noises became quiet and transparent and passed through my mind. The bell-ringing of the Master and other sounds were heard calmly and quietly. The queer roaring voices which had been annoying were now accepted with sympathy, and I sincerely hoped for the success of the other participants. In a word, a state of mind of forgiving all, sympathizing with all, and free from all bondage had been developed. I had not known of this before.

At the personal interview the Master explained that state of mind very kindly, and his words: "This is just the beginning of the proper training" appealed to me deeply. The thin, small figure of the Master seemed so noble and great. It was as if I could peer into the depth of the Master's personality a little. Breakfast was finished, and the next sitting began. Tears came no more. The feeling of surprise also was no longer there. The change in the basis of mind, however, was clear to me. At the talks during the recess and after the *sesshin* in the afternoon, I heard someone say that no one could have a sudden change in his mind and he could not believe in any "exaggerated" reports. I was wondering if there were large individual differences in susceptibility to inspiration. I just listened to their talks without any comment on my being so much moved.

I returned home next noon, leaving Tokyo by night train, passing through the Tokaido and the Chuo Lines. It was fortunate that the car was not crowded and I could maintain the state of mind developed during the *sesshin*. I attended my conference the following day, but I was able to keep my calmness through the conference.

Since then I have sat every morning and evening for more than two weeks, but I have found that I do not have enough time. During the five days of the training I sat with closed eyes, but now I sit with opened eyes, following the Master's advice. In the beginning it took about twenty minutes to enter the proper state of mind with opened eyes, but now it takes only ten minutes. If I continue the practice for one year, I believe that I shall be able to enter the supreme state immediately. I will make efforts to be able to act in everyday life with such a basis of mind.

The following are excerpts from the report of a contemporary American psychotherapist, Jack Huber, who had an intense Zen experience: *

The *Roshi* helped the students take their positions for meditation. Each person seated himself facing a side wall; half of the men were on one side of the room, the other half on the other side. The walls provoked no distraction; they were blank. The *Roshi* helped me get comfortable sitting on the floor. Two cushions were placed one on top of the other to raise my buttocks from the floor; my legs were crossed and my knees rested on the floor. My neck and back were completely straight. My back never became tired.

The *Roshi* arranged my hands very carefully. They were put in my lap, right fingers together and closed over the left, thumbnails touching and upright so that the nails were on top. The thumbs stayed in the prescribed position only if one were awake. As one dozed, they fell.

My eyes closed (I never mastered keeping my lids slightly open as is apparently the very advanced way), I began my meditation. The system had been described the night before and I had memorized the sounds I was to make and fix my mind on. We were to count in Japanese to three by breaking up the syllables of each number in this fashion:

HI-TO-TSU—U—U—U
HU-TO-TSU—U—U—U
MI-IT-TSU—U—U—U

The U was given an *oo* sound. There was nothing magic about Japanese numbers; we could have counted in any language, the numbers of which broke well into syllables. Each number was said in a normal breath and rhythmically, the U's continuing until the breath expired. A normal intake of air and then the number 2 (HU-TO-TSU), and then 3. After this, we were to return to the first number. The sound was to be in the throat, audible to oneself but inaudible to others.

Subsequent exercises were audible to everyone if one listened, but it was important at first that the room was as quiet as possible to aid in fixing the mind in the early stages.

* From Jack Huber, *Through an Eastern Window* (Boston: Houghton Mifflin Company, 1967). Earlier published under the title *Psychotherapy and Meditation* (London: Victor Gollancz, 1965).

Again with the minimum of English, the *Roshi* explained that we were to *listen* to the sound we were making in our throats—and we were to listen to nothing else.

I began my meditation. I listened to HI-TO-TSU-U-U-U, took a breath, and began to wonder what the function of counting was. I returned to listening and then began to think about why I was here. Back to counting and listening. And then more thoughts. What were other people doing? Were they having as difficult a time of it? Had I tipped the taxi driver sufficiently? Did the *Roshi* really know what he was doing? Could I stand going through this? My god, it was boring. My thumbs relaxed. I was falling off to sleep. MI-IT-TSU-U-U. Should that sound have been longer? No, that was a normal breath. Was I breathing the right way? Would the train continue to pass all day with its whistle and forceful shaking of the house? Did I only imagine the house shook? Back to counting and listening. HU-TO-TSU-U. MI-IT-TSU-U-U. I could not possibly keep my mind on it.

Bored, irritated, exhausted, I was interrupted by the sound of a bell in the room. I turned my head and discovered the *Roshi*; he held the bell and motioned for us to rise. Following him, we walked slowly in a circle, one behind the other, around the room. I guessed this was interjected to allow us to stretch and to allow some relief from an impossible task. About four times around the room, another tapping of the bell, and we sat down again.

I had perhaps never been so frustrated in my life. I tried to continue the counting and listening but I did no better with it. Again the frustration was interrupted by a bell. It was 11:30. We had been in the room over two hours now. We were to go downstairs one at a time for our private interview with the *Roshi*. Since the *Roshi* was already downstairs, I could have waited my turn by walking about the room or going out onto the balcony where there was a view over the rooftops of the suburb. With relief in sight, I decided to continue trying with my counting and listening until I was to go downstairs.

The private interview (*dokusan*) was held in our bedroom, which, of course, was free of furniture except for the low, round table. I had no idea about the ritual of the *dokusan* or what was to happen. We did not have any order of going in to see the *Roshi*; we went in according to who went downstairs first. I wondered how I could have a private interview when we did not speak the same language, and the interpreter was not to be here during the day.

I took off my slippers and walked into the bedroom. The *Roshi* was sitting on the floor in Japanese fashion, on his knees. I stood wondering what to do. He indicated that I should bow, go to my knees and bow again, this time all the way to the floor. As usual, his instructions were given with gestures and the few English words he knew. I realized this was a definite ritual; I assumed it was a Buddhist ritual of respect for one's teacher. I recall being somewhat amused by the deep bowing but I suppressed a smile. Oddly, I had the feeling that he was smiling but I did not know him well enough to be sure. I did not yet know that amusement might be shown about anything.

Having finished my bows, I sat upright again on my knees and looked at him. He looked at me for some time, seeming partly to try to read in my face what my reactions were and partly to be searching for English words.

With pencil in hand and tablet before him, he asked simply, "What per cent?" He pronounced it "*par* cent" and I never ceased in the next few days to find it amusing.

Somehow or other I knew that he was asking what per cent of time I was centered on the exercise. "Two," I replied. I was somewhat hesitant; I suppose I was waiting for criticism. He made a single mark in his tablet (I assume it was the number 2) and then looked up at me. He hesitated; he was searching for English words.

"Ah . . . you . . . try . . . try hard." He thought a second, nodded his head as if he were satisfied that he had said what he intended to say, and then smiled. He bowed his head; the *dokusan* was over.

Again he instructed me about the ritual. I bowed, got to my feet, bowed again, and backed my way to the door. The subsequent interviews followed the same ritual.

I returned to the room upstairs, seated myself on the floor, and continued my attempts to fix my mind on the counting. At about this time I began to notice severe pain in my knees—periodic sharpness and continuous burning, undoubtedly the result of having my knees bent for such long periods. I was distracted by it—only one of a thousand distractions, but a strong one. I would interrupt my attempts at centering my mind to rub my knees. I had no idea that the *Roshi* was observing me.

How painful will my knees become? Rubbing. HU-TO-TSU-U-U-U. Listening. What does the per cent have to do with it? What time is it? MI-IT-TSU-U-U-U. Listening. I'm sleepy. Counting. Not listening. I must really try to concentrate. Counting, listening. Counting, listening. I'm doing better at it. That is, of course, a distraction in itself. Counting, listening. Counting, listening. I nodded and pulled myself up with a start. I had fallen asleep.

I thought lunch would never come. Finally a metal gong was rung downstairs and all of us looked around. The *Roshi* was gone. We stretched, glanced at each other, smiled, arose, and went downstairs.

Lunch was taken with the same customs as breakfast. After lunch I again slept soundly for an hour and returned upstairs to my attempts at meditation. The afternoon went a little better, although my knees continued to hurt. When the afternoon *dokusan* came, I was able to estimate to the *Roshi* that I had been fixed on my listening about 10 per cent of the time.

The remainder of the afternoon was filled with feelings of anger, interest, frustration, determination, amusement—and what seemed to me to be a great deal of dozing.

Sometime during the afternoon I heard the *Roshi* talking to my roommate, who sat behind me facing the opposite wall. When we got up to relax I saw that he had taken to sitting in a chair.

The Japanese students began to include me in their glances about the room. Occasionally one of them would smile. At one point I risked a shake of the head at the young man sitting next to me, and said, "Hard!" He grinned and said, "Yes."

When I became unusually frustrated or bored, I was encouraged by the fact that other people around me were trying.

Finally dinner was announced. After dinner we again slept for an hour. I was becoming accustomed to the naps. I never had trouble going directly to sleep.

When I got up from my nap and passed through the main room, I met the young Japanese woman.

"How is the *zazen*?" she asked.

I found that I did not want to talk and I answered briefly that it was hard. She came upstairs with the group, and I saw the *Roshi* instructing her in *zazen*.

The evening went a little better but I was still having trouble with my knees and a thousand distracting thoughts.

*Zazen* over, we went downstairs and the *Roshi* spoke to the interpreter, my roommate, and me. This happened periodically; I assumed he was giving us part of what he gave the Japanese students at another time. During these infrequent and short conversations he spoke about Zen.

To his statements I could give only highly personal associations but I was never asked to state them. Even if we could have communicated in a common language, I do not believe he would have pursued my associations to help me analyze them. This was left for me to do as I chose. He and I were in no way "in this thing together," as therapist and patient are apt to be. We were mutually sharing neither a review of my daily activities and thoughts nor a recounting of my past. He pointed to vague goals, the content of which I would have to fill in for myself. He instructed me in meditation; he contributed to my comfort and the elimination of distraction. The rest was up to me.

The interpreter got up to go and I asked if she would like me to walk home with her. She seemed pleased but doubtful. She turned to the *Roshi* in a casual way, spoke, turned back to me, and smiled.

"You are not to leave the house." She seemed to expect his reply; I should have known he would try to prevent any distraction for me.

When she left, I returned to my room. As I took out our bedding from the porch closet, my roommate began to talk about *zazen*—Did I see its similarity to one psychological principle or another? How was I doing? What did I think the function of counting was?

I was sure the rule against unnecessary conversation was crucial to the training and I wanted to follow the rules. I answered his questions perfunctorily, irritated. This occurred a few more times in the next day or two until I mustered the courage to tell him that I did not want to break the rule against unnecessary conversation. I gradually ignored his overtures to conversation and he became more silent. Perhaps there is something important about this. Before he became almost completely silent, my irritation reached a high point. I recall thinking I would push myself to follow all the rules despite distractions provided by my roommate. He was unquestionably a goad, perhaps an important one.

I turned off the light and got into my bed on the floor.

"The hardest work I have ever done," I remember thinking as I fell off to sleep.

The second morning of *zazen* went much the same as the previous afternoon. I dozed a great deal, partly, I believe, because I was still tired from my long trip to Japan, and partly because I was bored, frustrated, and discouraged. My knees continued to hurt.

I was trying hard to attend to my counting and listening when suddenly I felt a tap on the shoulder. I was startled and I turned to find the *Roshi* beside me. Again he looked at me in an apparent effort to find English words.

"Floor . . . for Japanese. No Westerner." It must have been apparent to him that I was determined to learn to sit on the floor, but he made the decision for me that the distraction was too great. He motioned for me to get up. He went to a closet and brought back a chair on which he placed a cushion; he also put a cushion on the floor for my feet. He indicated that I should sit on the chair. I again sat very straight; my back was supported by the chair. He put my feet on the cushion. My hands were in the usual position, thumbnails up to check my dozing. He stayed beside me until he was satisfied that I was comfortable. My knees stopped hurting immediately.

I began to do a little ·better at listening to my counting and cutting away other thoughts. Despite the dozing, I was able to report a some-what higher *par* cent of time in attending to my counting and listening.

After lunch and the usual nap we returned to the room upstairs. This time as we seated ourselves and were about to begin counting, the *Roshi* interrupted. He announced that we had finished the first phase of our training. He again spoke in Japanese and then would attempt an English word or two of translation. Gradually and hesitatingly the young man who sat next to me began to translate. He seemed to feel totally inadequate to the task, but there was a task to be done and he did it. Actually his English was quite fluent; the *Roshi* had obviously not been aware of this.

The *Roshi* explained that we were to produce a single syllable—MU (the U, as usual, being an *oo* sound)—in a rhythmic fashion. On each expulsion of breath we were to produce MU-U-U-U until the air was out. Again it was to be throaty and barely audible. *Mu* (or *Moo*), I realized later in rereading the *Roshi*'s pamphlet, means "nothing."

I welcomed the variation. It later occurred to me that repetition produces boredom, and boredom brings distraction. I assume the *Roshi* uses variation to cut down the time required to produce complete centering of attention. I must add that I *think* this was a planned part of his method; I had neither opportunity nor language to discuss his methods with him. Furthermore I did not want to intellectualize at the moment.

My dozing was becoming less frequent. I was clearly doing better and by the time the afternoon *dokusan* came, I was able to report a still higher per cent of attention.

Sometime in the afternoon I had two peculiar sensations. I do not know how long either of them lasted. I first noted a feeling that my body was becoming heavier and heavier. The feeling finally disappeared. Later in the afternoon I had the feeling that my body was growing larger and larger, almost as if it were filling the room. I had experienced the latter feeling only at one other time in my life; I recall that during childhood it occurred on several occasions before sleep.

Much later in the afternoon I had a striking insight. In the middle of my sound-listening periods I realized that no matter what thought or feeling I had, it passed on. It either never recurred or never recurred in quite the same way. I realized that nothing will ever come again as it comes to me at any one moment. Nothing exists but what exists now.

Later I recognized this as one of the cardinal Buddhist principles—the impermanence of all things, the constancy of change.

These were not totally new thoughts to me, but in the intensity of observation of what went through my mind I experienced the idea as I

207

never had before. At each moment only sound, then only listening, then breathing, then sound, then listening. Then distraction—a face, a thought, a feeling, an idea. Anger, frustration. Would the distraction last? That was the question. Will it last? And then it struck me. Will what last? Was this not the question I was always asking myself? Will this pleasure last? This anxiety? What can I do to make it go away? Always the avoidance of what is going on at the present moment. Always the concern about the future, tomorrow, next year, the next moment. What next moment?

Oddly my reaction to the ringing of the bell, this time the dinner bell, had changed; I was not quite as relieved as I had been previously. I was pleased to stop but the eagerness to stop *zazen* was no longer there.

I believe it was that afternoon when my roommate began leaving the house. When I came downstairs to eat, I found him cleanly shaven with the smooth, shaved look that only a barber produces. He explained that he had had to have a real shave and had gone into the neighboring village. During the following days he continued to visit the village during *zazen* and would usually bring back candy or ice cream for all of us. In the middle of *zazen* I would hear soft moving, would look up, and see him tiptoeing out of the room. I never knew how much time he spent away from the center.

I was becoming quieter. I was generally less distracted, more centered on what I was doing at any one moment.

After dinner and sleep the interpreter again came to sit in *zazen* with us. When the evening session was over, I spoke briefly with her. These brief conversations allowed by the *Roshi* were, I believe, intended as an opportunity for us to ask the interpreter anything we wished since his own lack of English did not allow him fully to know what was on our minds.

When she came up to me, she was obviously troubled.

"You're changed," she said. "You look different."

I had no reply to this and did not really want to think about it. I was not willing to attend to framing a reply. I knew also that the minute I began explaining or labeling, my attention would shift to the explanation and away from the quiet centering of attention.

There was no suggestion of my walking home with her; the *Roshi* had made that quite clear to both of us the previous evening.

Quietly I went to bed. The interpreter was right. I was different. I suppose the word is undistracted. I was centered on what I was doing, although I did not think of that at the time. As I looked back on it, I was neither happy nor unhappy; I had dropped even that kind of labeling.

I now seldom paid attention to the train or its whistle. At those times when I did notice it, I listened to it. No longer concerned about its distracting me, I quite enjoyed it when I heard it.

The morning of the third day we continued with the same exercise. Lunch was quiet, and then the nap came. In the afternoon the *Roshi* gave us a new exercise, our third. We were to make a single sound with no rhythm or pulsations. The sound was the previous one, MU. It was to be made with power. As demonstrated for us, the U part of the sound was almost like a GR-R or growl. We were no longer to hold in the sound; it was to be audible. It seemed simply like another exercise to me; not until later that evening did I realize the quality of strength and power and even the strangeness of it to an observer.

I realized now that each of the exercises required considerable attention even to do them accurately. The strong MU sound captured my attention and interest. One could hardly make the sound strongly without devoting attention to it. Actually this was true of all the exercises. At first I was somewhat distracted by the loud noise of the five of us making the strong MU sound. The distractions, however, soon disappeared when I began making the sound as prescribed and listening to it. When I went for the afternoon *dokusan* my *par* cent had risen to an estimate of about 50 per cent of the time centered on the task.

By this time I was no longer thinking about whether or not I was doing well. I had stopped making comparisons of this kind; I had become free from that kind of distraction.

Dinner and nap came and went. I returned to the upstairs room and noted in passing through the door that our interpreter was speaking softly to the *Roshi*. Totally involved with *zazen* by this time, I did nothing but note the fact that she was there.

The last bell was rung and I went downstairs. There in the central room was our interpreter; she walked toward me immediately as I entered the room. She was distraught.

"Please forgive me," she said, "I'm completely shaken. I've never seen anything like that—to come into a room and see all of you doing that. I just can't take it. I don't think I can stay. I feel terrible about leaving you both this way, but I don't think I shall be able to come back. Please forgive me."

My first reaction was anger. I needed her to translate in the *dokusan*. I wondered if I would have to stop *zazen*. I had forgotten that my *zazen* neighbor could speak some English; he seemed so shy that I did not think of him as an interpreter. I could do nothing about convincing her to stay. I said simply that she must not worry about us. She left seemingly undecided about returning. The *Roshi* seemed undisturbed by the incident.

209

That was the end of the third day. I knew I would continue with or without her interpreting.

The routines—eating, *dokusan*, naps—continued as I have described them. My relationship with the group had changed, however. At first I had felt very much an outsider, a Western curiosity; now I felt I was a part of the group. In particular, the boy sitting next to me in *zazen* became friendly. I would occasionally have to ask him for a clarification of the *Roshi*'s instructions; he seemed always to want to help me. I was impressed by his unwavering attention to the exercises; his presence more than any of the others' often encouraged me when I felt like resting.

Sometime during the fourth day a new person came. At one point I looked up and saw a man of about twenty-five sitting on the floor near the balcony, a short distance from the rest of us. He sat in full lotus position, legs crossed and upturned feet resting on the opposite thighs. I remember thinking he must be a former student because of the lotus posture; it is a difficult position to assume and to maintain. He never talked to anyone. I did not know then that his presence would be crucial to me the next day.

I thought no more about the interpreter. It never again crossed my mind that I might have to quit for any reason.

Early that day, the fourth day, the *Roshi* gave us a totally new exercise. Holding a piece of paper and a pen in his hands, the *Roshi* called us together to sit near him. He drew a simple picture of a short funnel with a long beak. He explained that we were to produce a prolonged MU sound (the beak) and then gradually increase the force of the sound (the funnel) to a crescendo. The beak part was to be about two-thirds of an exhalation; the funnel the remaining one-third. During the latter one-third the face, shoulder, and arm muscles were to be strongly tensed. The increase in volume was to be steady, not wobbly or varying up and down. We were instructed to listen to the prolonged MU but *not* to the last one-third. Our hands were to shift position drastically; each was made into a fist and placed on the thigh.

While I listened to the description intently and seriously, I found it amusing. It was ludicrously ferocious. I smiled openly. I was beginning to learn that there was nothing so sacred or serious that I could not express my amusement about it.

We followed the exercise all day. I do not remember hearing the train at all. I was occasionally aware of the loud sounds made by the other students around me, but this awareness was momentary. Moments of my life, thoughts, faces, uncomplicated sexual images devoid of fantasy or story quality came briefly and went.

During the morning I felt another peculiar bodily sensation. My body seemed to rise, while the floor, which I could not see since my eyes were closed, was sinking. I have no idea how long or how often I felt this. The feeling was, I recall, neither frightening, pleasant, nor unpleasant; it simply occurred and I observed it.

The afternoon session was without incident. I reported in my *dokusan* that I was focused about 80 per cent of the time.

That evening the interpreter returned. She had apparently thought over the situation and had resolved her conflict. I was pleased. I liked her and needed her assistance; my attention to the exercises had reached a point where I wanted an easy exchange in the *dokusan* with no language barriers.

The interpreter and I went in to see the *Roshi* when my turn came. She sat on the floor in Japanese fashion with her legs behind her; she placed herself to the side where she was unobtrusive. When she was there to translate, the *Roshi* could express himself fully; he did not have to look for English words as he did when he and I were alone.

I faced the *Roshi*. As always, he looked directly at me.

"What did you hear during the last one-third of the exercise?" he asked.

"Nothing," I replied.

"What did you feel?"

"Strength," I found myself replying. I had not thought of the word before he asked me.

And then he said, "That is your nature. That has always been there to come out."

This was the *dokusan* as I remember it.

It was the fifth day. I took no particular cognizance of the fact. Looking back on it, I was eating more slowly; my movements were slower. What I did not choose to give my attention to (the train and its whistle, the loud noises of the current *zazen* exercise) I was unaware of. What I chose to attend to (*zazen*, meals, gestures, brushing my teeth, the thoughts and images passing through my mind) I was acutely aware of.

I arose from bed, washed and shaved, practiced *zazen*, and came to breakfast. There the *Roshi* made an announcement. I find it among my papers, written in the handwriting of my *zazen* neighbor, the serious boy. I must have asked him to do this but I do not recall the circumstances.

The translation was this: "The last practice of today's *zazen* is just equal to this four days' practice. It is just for this last *zazen* that we have held this Zen gathering or *sesshin* for these five days. You must not neglect this last chance. Please do your best and increase your efforts."

In this announcement the *Roshi* did not say *why* one should do his best or increase his efforts. He did not say what the last chance was for.

I believe my only goal at the time was to center my attention as I had been instructed. It was my last chance to do just that.

After the usual morning nap, we went upstairs. Seated again, we were given the fifth exercise, a variation of the previous one. This time the beak of the funnel, the steady MU sound, was to be short and was to be followed by a longer U of rising intensity to a crescendo. The ending of the sound was to be cut off sharply. We were to listen to it all.

The time must have been eight o'clock.

For some time there was little variation in my previous high degree of centering my mind. The new exercise involved no great variation, although there was the longer intense sound and muscular tension, the sharp ending, and total listening.

Gradually I must have become completely focused on the exercise. I do not think that anything else went through my mind.

And then—it was late in the morning—a white, clear screen came before my eyes. In front of the screen passed or, rather, floated simple images—faces, objects. I have no clear recollection of the images. A rush of feeling came over me.

I burst into tears; the tears became quiet sobbing.

I do not remember at what point I had stopped the exercise.

I can state my feeling but I am not sure I can communicate it with any real meaning. I would like not to be mysterious; I would like to communicate it clearly, at the same time knowing that it may be impossible.

My feeling was that I was seeing something of great importance, as if everything fitted together for the first time. What had all my life struggles been about? Things were very clear and very simple.

I do not know how long I sat sobbing. Someone was at my side. He had his hand on my shoulder. It was my *zazen* neighbor, the boy who sat next to me. He took my arm and I arose from my chair; we walked slowly together down the stairs. The new member of our group, the quiet man who had come on the fourth day, was downstairs waiting for me. I realized he was to be my interpreter; he had apparently volunteered.

He and I entered the *dokusan* room. I bowed and sat on my knees facing the *Roshi* as usual; I was still sobbing. The quiet man sat to the side. He translated easily and fluently. As he translated he looked directly at me. He now seemed warm, interested, and to want me to understand clearly what was being said to me.

"You have seen *kensho*," the *Roshi* said simply. I was aware that *kensho* meant a glimpse into one's own nature.

I did not reply. There was a long pause. The *Roshi* said something in Japanese and pointed in front of him. I looked at the quiet man.

"He says to come close to him." On my knees I slid closer.

"No, closer." I slid forward. Our knees were almost touching.

He opened his mouth quickly and burst forth with a loud sound, like a sharp "Ah!" I started. He looked at me.

"What did you feel?" he asked.

"Surprise."

"And after that?"

"Nothing."

A pause. My ankles were hurting. I rubbed them.

"Are you feeling well?" he asked.

"Yes. Only my ankles hurt." I had not become accustomed to sitting on my turned-under feet in the Japanese way.

"Get up and walk about."

I rose and walked about, rotating my feet to move my aching ankles. Relieved, I returned to my sitting position.

The *Roshi* looked at the place where I had walked.

"Are you able to see the footsteps?" the *Roshi* asked.

"No."

He nodded his head. "They were not there before and are not there now. There was nothing in your life before and nothing in the future, only—" and he burst forth again with "Ah!" This time I did not start. I looked at him. There was a pause.

"What do you think of life and death?" he asked.

"I don't think about it," I answered honestly.

"There is no life and death, no me, no you, only—" and he almost shook the room with "Ah!"

Again he looked at me intently for a few seconds and said, "Now you must rest. No more *zazen*. Go up the stairs and sit quietly."

I bowed to him and to the quiet man and arose. My crying had stopped. I returned upstairs and sat quietly in my usual place until lunch. I was aware of the sounds around me. People continued with *zazen*, the train passed and whistled, occasionally feet moved across the room. I listened, thoughts and faces and ideas passed through my mind, and I looked and listened.

Lunch was silent. I slept for an hour and returned to the room upstairs. I sat quietly again. The afternoon *dokusan* hour came. I waited my turn and went downstairs. Again the quiet man was with me. I entered the room, bowed, and smiled at the *Roshi*. He returned my smile and then looked serious.

"You have succeeded in Zen," he said.

I did not reply.

"Did you come here with serious personal problems?" he asked.

"No." He looked at me, paused a few seconds, and nodded.

"You must continue *zazen* or this will become only a distantly re-membered experience. I know you are busy in the big city. But in your busy life, sit in *zazen* each day if only for five minutes. Begin with the third stage. But do not begin until one hour after eating. Otherwise, it will be bad for your stomach. Do not do the exercise from the stomach, but from below the navel—as if you were pushing the navel up."

I nodded.

"And remember. Do only that which is right—for a time and a place and a situation. Reject any action that has not these three conditions. And do good things for your fellow men."

He again looked intently at me for a few seconds and nodded. And then he smiled.

Finally only the interpreter was left. I asked her if I might walk home with her. She turned to the *Roshi*, who did not need a translation; he nodded.

She and I walked again through the suburban streets. How long had it been since I was outside?

There was no mist; it was not raining. The street lights shone brightly on the houses. When we arrived at her house, she asked me to come in to meet her family. They were congregated in the living room when we came inside; I realized my visit had been planned. I sat for a few minutes but I did not really want to be there at the moment, making conversa-tion. I felt Zen would be a topic and eventually it came up. I turned it away, unable at the moment to communicate anything about my experi-ence.

I returned to the *Roshi*'s house and the *okusan* helped me prepare my bed.

I was alone for the first time in five days . . . was it only five days? I took off my clothes and lay down in my floor bed. I propped myself up on my elbow and looked out of the glass doors of the porch. The moon was shining and I could see the small trees outside. The train passed and looked alive with light; people sat in the coaches reading their newspapers. The whistle blew, the house shook, and I found myself smiling.

The next morning I arose casually and washed and shaved. The three of us had breakfast; for the first time the *Roshi*'s wife sat at the table. I had my eggs, toast made in the toaster, jam, butter, and instant coffee. When breakfast was over, I put my bowls and napkins together in the usual way as if I would be there to eat again. I helped the *Roshi*'s wife clear the table. She would not let me help her clean the dishes.

Somehow or other the *Roshi* and I communicated quite easily. I spoke simply and slowly and the *Roshi* searched his limited vocabulary to tell me how he established the school, how he wished to come for a visit to the United States to instruct in *zazen*, and other things I do not recall now. When spoken words were difficult to understand, we wrote them and sometimes even drew pictures to communicate what we wished to say to each other. The *Roshi* translated to his wife what we said.

After breakfast the *Roshi* went into my bedroom and sat on the floor by the low table. I followed him. I do not remember any special preparation for what followed. The *Roshi*'s wife brought red and black inks, brushes, seals, and some heavy picture-size white cards with gold edges. While she and I watched, the *Roshi* began painting calligraphy on the cards with his Japanese brush and black ink. He handed a card to me, obviously a gift. The only part of it I could read was my own name. I asked him to translate and I wrote as he attempted to translate with his limited vocabulary. What I wrote down was the following:

Congratulations. J. T. Huber has seen unconditional nature.
In nothing there is no end.
There is flower and moon and high house. Many clouds in the sky.
All people cool. [He explained that people in pain are "hot."]
 Moreover, with effort comes high understanding.

He used two seals, pushed them one by one into a pad of red ink, and carefully made impressions on the card. One of the seals carried his own name. The other impressed the words "No moving mind."

I was amused by the congratulations. I bowed my head and thanked him.

The *Roshi*'s wife brought out a painting she had done—a ferocious and amusing male face on a large sheet of fine rice paper; the size was appropriate for a scroll. This, too, was a gift for me.

The *okusan* put away the paints and the *Roshi* said we must take photographs. I indicated that I wanted to change clothes for the occasion. He seemed pleased, and the two of them went to collect the camera equipment. I changed from my usual casual clothes to my dark blue suit, white shirt, and tie. . . .

I put my bags down, put my hands together in the usual position, and bowed. The *Roshi* and his wife did the same. When I raised my head, they both looked directly into my eyes. All three of us smiled. I lifted my bags and began walking away. I looked back every few yards. They bowed each time and smiled. When I reached the main road I looked back for the last time; they bowed and then waved. They walked slowly toward their home, looking back and waving until they were out of sight. I can still see them.

How can I tell the next part of the story? Perhaps we do not describe things well when we have lived them fully; could that odd twist be true? Perhaps all of us find it difficult to communicate our feelings when we are really feeling something deeply, when we are really seeing things clearly. How does one tell another person what it is like to be fully aware of what one is feeling, what it is like to love, to feel anger?

I returned to Tokyo and began sightseeing, going to restaurants, meeting people. Then a visit to Kyoto with its tiny pebbled gardens, its small summer residence of the Emperor, its ancient Shogun's palace, its countless huge temples.

I was seeing it all as if I had never seen it before. I seemed almost to have a new pair of eyes, new ears, new abilities to taste and smell and feel. I had learned to give my full attention to whatever I was doing at any one moment and I wondered if I had ever really done this before.

Gradually I began to see I was eating when I was hungry, not when it was "time to eat." I began to eat what I wanted to eat, not because it was placed before me, because others were eating, because we must have three good meals a day. I was reminded of the psychological study of small children who, allowed to eat what they wanted to eat, chose in the end a balanced diet. The extra weight I had picked up in my early forties began to disappear and has not reappeared. Perhaps we have no "weight problem" if we eat when we are hungry and eat what we really want to eat.

I began to feel I had never really tasted things before. I ate less, drank less, and enjoyed both experiences more. Even being with people became a new kind of experience to me. I had always been gregarious—and often undiscriminating. Now I chose *whom* I wanted to be with and now I was *with* them. I was seeing and choosing what I wanted to do—speak with someone, drink coffee, read a book.

I saw what I was doing as if I had never seen it before. And the pleasures I found in it all were something I could not have imagined.

## PRACTICAL WISDOM

We add here two brief notes, both written by our consultant Dr. Tadashi Tsushima, comparing and contrasting Japanese and Western thought. The first of these describes the functional relations of mind and body: the influence of physical conditions upon mental activity and the influence of psychological processes upon general health. The second relates to that exquisite concern with aesthetic form so characteristic of every Japanese temple, or flower display, or drama, extending over the whole domain from etiquette to self-control. The essay on *iki* illustrates the subtlety and charm of the Japanese analysis of things both physical and psychological, and will bring to a conclusion these brief notes on Japanese psychology.

### Psychosomatic Medicine—Some Wisdom in the Japanese Classics*

The intimate relationship between the mind and the body, and what we now call psychosomatic medicine, was recognized long ago in Japan. Ekiken Kaibara (1630–1714), for example, in his most famous book, *Yojokun Lessons for Good Health,* stated that the first and most important lesson for good health was to control or get rid of things harmful to the body—namely, desires and unhealthy external conditions. By desires he meant those for food, sex, sleep, and talking; he also included the feelings of joy, anger, enjoyment, anxiety, worry, sorrow, and surprise. By unhealthy external conditions, he meant too much heat or cold. When one becomes angry, *ki* goes too high; when one is pleased too much, *ki* loses strength; when one is sad, *ki* disappears. *Ki*, which is an old traditional concept in China and Japan, represents mental force or psychological energy. When one is afraid, *ki* does not move. Ekiken believed that most diseases are caused by inadequate conditions of *ki*.

In order to prevent disease, one should enrich *ki* (by maintaining a peaceful mind, controlling anger and lust, and lessening anxiety). Ekiken believed that tolerance and patience were the most important mental factors for good health and a long life. He also gave practical advice concerning the activities of daily life, and discussed the proper use of alcoholic beverages and sex. Drinking *sake* in moderate amounts, for example, is very good for the health both physically and mentally; however, nothing could be worse than drinking too much *sake*. One should also control one's sexual life; an average man should not ejaculate more than once a week in his thirties, more than once in two weeks in his forties, more than once in four weeks in his fifties, and after sixty he should not ejaculate at all. In addition, Ekiken recommended a method of sexual contact that does not culminate in ejaculation.

Genpaku Sugita, who translated *Tafel Anatomia* into Japanese,

* Written for this volume.

stressed two points for mental as well as physical health—not to worry about past misfortune and not to worry much about the future. If one fails to understand these two points, many diseases may result. He thought that conscious will and decision-making are the most important factors for realizing these two points.

Thus these two important Japanese scholars of olden times maintained that a peaceful mind is the most essential thing for health. Their ideas would seem to anticipate some of the stress theories of contemporary psychosomatic medicine.

### Iki—A Japanese Concept of Beauty*

Concerning the aesthetic consciousness of the Japanese, various explanations and interpretations have been made, and one of the best was given by Shuzo Kuki (1881–1941), professor of the history of Western philosophy at Kyoto University. In his book *The Structure of Iki* (1930), he explained the special character of the Japanese aesthetic consciousness by analyzing *iki*, which was the concept of beauty that flourished in the eighteenth and the earlier part of the nineteenth century in Japan. Though *iki* has somewhat lost its influence among the Westernized Japanese of today, it is still one of the most significant factors of Japanese traditional beauty, and Kuki's analysis has not lost its importance.

He begins his analysis by distinguishing *iki* from its Western equivalents, such as *chic, coquet,* and *raffiné*. According to Professor Kuki, *chic*, which means delicacy or excellence in taste, subsumes both *iki* and elegance, and is more comprehensive than *iki*. *Coquet* is a narrower concept than *iki*, which comprises other qualities. *Raffiné* is also narrower than *iki*, and lacks the essential elements of *iki*. Thus, he concludes that *iki* is particular to the Japanese, and maintains that *iki* should not be studied through ideation (in Husserl's sense) but through hermeneutics. From this hermeneutic standpoint, he maintains that the comprehension of *iki* as a conscious phenomenon must precede the apprehension of objective expression of *iki*.

In order to comprehend *iki* as a conscious phenomenon, it is necessary to explicate the intensive and the extensive structures of *iki*. Professor Kuki points out three characteristics of *iki*: (1) coquetry, (2) high spirit, and (3) resignation. (1) *Iki* is found in some love affairs which depend upon coquetry. "Coquetry is," says Professor Kuki, "a dualistic attitude in which one establishes a relationship between oneself and the opposite sex." Coquetry aims to charm the opposite sex, and is ex-

* From Tadashi Tsushima, "Iki," *Psychologia*, VI, Nos. 1–2 (March–June, 1963), 71–73.

tinguished when the end has been achieved. Duality is the fundamental element of *iki*. (2) Though *iki* may be regarded as coquetry to some extent, it also contains a kind of resistance to the opposite sex. *Iki* implies not only compliance with the other sex, but also dignity derived from high pride or spirit that originates in *bushido* (Japanese chivalry). (3) The third mark of *iki* is resignation or unattachment which is attained through adversity. It goes without saying that this resignation has a close connection with Buddhist pessimism. These three characteristics of *iki* may, at first sight, seem to contradict one another. But the fundamental attribute of coquetry has a dualistic possibility, which is sharpened by idealistic high spirit or pride. Besides, coquetry can exist so long as its end is not attained, and therefore, resignation to the end is consistent with coquetry. In short, *iki* may be considered as coquetry spiritualized by high pride and resignation, or by idealism and negation of reality.

In regard to the extensive structure of *iki*, Professor Kuki compares *iki* with similar concepts of beauty, such as *johin, hade,* and *shibumi*. These concepts, in turn, have their antonyms, such as *gehin, jimi,* and *amami*. Of these aesthetic concepts, *johin* and *hade* are based on general human nature, whereas *iki* and *shibumi* are based on intersexual particularity. (1) *Johin* and *gehin*. These imply "to be better in quality" and "to be worse in quality," respectively. They are found in human affairs, especially in taste. Thus *johin* means gentility or elegance; *gehin* means vulgarity. *Johin* shares excellence of taste with *iki*, but it does not contain coquetry, as does *gehin*. *Gehin* is often connected with coquetry, though it is far from the excellence of *iki*. (2) *Hade* and *jimi*. *Hade* (gorgeousness) means intense self-assertion and *jimi* (quiet taste) means little self-assertion. Neither is related to value in themselves. *Hade* is sometimes brought into a closer relation with coquetry, but it is inconsistent with resignation, which is included in *iki*. *Jimi*, of course, has nothing to do with coquetry. It can develop into resignation, however. (3) *Iki* and *yabo*. These are related to value, or based on judgment of value. *Yabo*, the contrary of *iki*, is etymologically derived from *yafu* (boor) and signifies boorishness. (4) *Shibumi* and *amami*. These originally stand for astringent taste and sweet taste, respectively; but in its figurative use, *amami* has a close connection with love affairs, and the normal states of love affairs are described with the words connected with *amami; shibumi* (taste of *shibui*) is, of course, the opposite of *amami* in this sense. *Shibui* is not the contrary of *hade*, but the negative of *amami* (sweet). *Iki* is intermediate between *amami* and *shibumi*. *Iki*, which contains resignation, is, in a sense, the negation of *amami*, and at the extremity of this negation there appears *shibumi*.

The extensive structure of *iki* may be summed up as follows:

| | | |
|---|---|---|
| based on general human nature | in itself (related to value) | *johin*(valued) |
| | | *gehin*(unvalued) |
| | to others (not related to value) | *hade*(positive) |
| | | *jimi*(negative) |
| based on intersexual particularity | in itself (related to value) | *iki*(valued) |
| | | *yabo*(unvalued) |
| | to others (not related to value) | *amami*(positive) |
| | | *shibumi*(negative) |

The extensive structure of *iki* can also be diagrammed in the form of a cube. The bottom and the top of the cube, respectively, show human generality or general human nature and intersexual particularity; the eight corners of the cube represent the eight concepts. Though all the corners (the eight concepts) contrast with one another, the two diagonal corners on the top or the bottom contrast most conspicuously. The rectangle O-P-*iki-johin* represents value; the rectangle O-P-*yabo-gehin* represents no value; the rectangle O-P-*amami-hade* represents positivity; and the rectangle O-P-*shibumi-jimi* represents negativity.

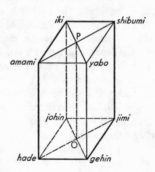

Other aesthetic concepts can be located in this cube. *Sabi* (antique-looking and chaste) is represented by the pentahedron whose bases are the triangle P-*iki-shibumi* and the triangle O-*johin-jimi*. *Miyabi* (urbaneness) is located in the tetrahedron whose apices are *johin, jimi, shibumi,* and O. *Aji* (piquant or smart) is represented by the triangle formed by *amami, iki,* and *shibumi*. *Otsu* (smartish but unusual) is situated in the tetrahedron bounded by *amami, iki, shibumi,* and *gehin*.

*Iki* can be expressed either physically or artistically. The main forms of the physical expression of *iki* are: (1) to have a good carriage or bearing, (2) to wear a transparent kimono, and (3) relaxation of the countenance. These expressions suggest coquetry and its restraint, which allude to high spirit and resignation.

The artistic expressions of *iki* are found in architecture and music. The form of the design which can be the expression of *iki* is stripes, especially vertical stripes, since they manifestly represent duality, which is the fundamental element of *iki*. As for the color of the design, gray, brown, and blue are regarded as the expression of *iki*. Gray, which suggests resignation or idealistic negation of reality, can express *iki* when it is combined with stripes that represent duality or coquetry. Brown, the darkened color of gorgeous red, orange, or yellow, and blue, the main cold color, suggest idealistic negation of reality. The architectural presentation of *iki* is found in a *chaya* house (an old-style Japanese restaurant or inn). By using contrasting building materials (such as wood and bamboo), and adopting an asymmetrical interior structure, duality is expressed; while the color (gray, brown, and blue) of the materials and the indirect illumination allude to idealistic negation of reality. In music, *iki* is expressed by somewhat ill-balanced melodies and rhythms, such as the change of the tone and the discord of the rhythm between a song and its accompaniment. These ill-balanced melodies and rhythms allude to duality. But the change is limited within three quarters of the whole tone, and the discord within a quarter of rhythm. These limitations are nothing but the objectification of idealistic negation of reality. In short, *iki* is objectively expressed through duality and its restraint.

Though in present-day Japan *iki* is not found as often as *shibui*, it still has some important meanings to offer in some kinds of drama and in various kinds of visual designs.

# A BACKWARD GLANCE

In introducing this volume, we made a good deal of the geographical setting within which Asian cultures have developed and the broad cultural matrix within which philosophies and psychologies arose. Now that our brief survey is over, and the reader has some glimmering of the kinds of psychological teachings which the great Asian philosophies developed, we shall offer a few more tentative guesses about the relation between Asian cultures and Asian psychologies.

If a speculative note be allowed to prevail, one may ask first about the *family system* and interpersonal relations in general as a fact predisposing to one type or another of psychology, whether in the West or in the East. It has often been noted that psychology of the immediate biological family—father, mother, and children—as developed by Freud in his book *Group Psychology and Analysis of the Ego*, offers a useful clue to the structure of the religious life and of religious concepts. The deity, for example, is usually conceived, in the West, as a paternal figure to whom male devotees direct their filial attention. There is, however, a need for a balancing emphasis upon femininity, as in the cult of female figures of divine or near-divine status. In religious orders there are brothers,

223

sisters, mothers superior, and so forth. The devotional life is organized in some ways around the themes of love and fear as they occur in the family. It has been noted in the same connection that this central role of the biological family does not appear to any great extent in the religions and philosophies of Asia. On the contrary, personal deities are seldom of absolute authority. Rather, they are secondary to vast, impersonal laws and principles, and when they appear, none of them has absolute power over the rest. There is, moreover, a group responsibility; so to speak, a social life, social authority, a precedence pattern in the Pantheon. The Pantheon of personal deities is in itself only a part of a vast schema of cosmic forces, many of which are semi-personal or wholly impersonal.

Now the question arises whether we might not properly assume a corresponding kind of religious philosophy as we turn to India, China, and Japan, where there is not a sharp separation of the biological or nuclear family from the extended family. These societies possess a rich system of brother-sister-aunt-uncle-grandparent-and-cousin relationships in the extended family. Such institutions give rise to philosophical and psychological systems much less sharply anchored upon individual existence, and upon the individual love and fear directed to a personal cosmic monarch. We might, in fact, expect to find in these Asian societies, with their extended-family systems, psychologies much less highly individualized, with less sharply defined individual personalities and status competition one with another.

The issue cannot be neatly resolved by historical data. Yet it is of interest to note that with the development of the family and caste system in India, there is the development of a psychology of caste, and of clan, in which individuality is hardly ever mentioned. It is rather extraordinary that we do not even have names for most of the Indian philosophers (whereas among the Greeks scores of philosophers' names are known). Every now and then an Indian philosopher with great force of character manages to battle against a great tradition and assert his name, but then for centuries his followers remain unspecified. Duty to father and mother, to husband and wife, is indeed constantly emphasized, but if one looks more closely, one finds that these are aspects of broader and more complicated social responsibilities. Both in the philosophy and in the great epic poems of India, the law of the extended family and the nation takes precedence over personal obligations. Indeed, the great Ram himself, who gave up a kingdom at his father's behest, and his devoted wife Sita, who clings to him rather than yielding to various supernatural claims and counter-pressures, are quite evidently carrying out their duty to a cosmic scheme. The point must not be pressed too far. But it looks as if the loneliness of the individual religious practitioner who follows the Upanishads or the ways of yoga is freeing himself from a social system as well as from a world of woe, and that he is doing this in retreat from all human relationships. He does not look for a reunion with the divine beyond the grave, as do both Christian and Moslem worshippers; rather, he is pursuing a process by which his own severe, pure, and eternal individuality will be set free in order to be absorbed ultimately into the nameless, but perfect, bliss which is available for those whose social existence has been purified and stripped bare, down to the ultimate pure reality of a selfhood which is no longer an individual self but only a divine spark.

When Buddhism, as a representative of a highly evolved Indian system of thought, was imported into China, it took with it very little of this mystical cultivation of the individual, but a great deal of the sense of social responsibility which Buddha's ethical teaching had established. It was partly because Buddhism often taught absorption in the eternal that some of the emperors were afraid of it, and it was precisely because Buddhism also taught responsibility to the social group that it finally fitted quite well into the ethical system—the system of family, group, and state relationships which was dominant in China. Neither in China nor in India, however, did a psychological system arise which dealt concretely with the specific problems of specific husbands, wives, fathers, children, as was characteristic of the Judeo-Christian applied psychology of the West.

To a considerable degree this held also for Japan, in which likewise the extended family is dominant. It is well known that from the time of the Samurai (the great warriors of the medieval period)

225

onward the Japanese have cultivated selfless devotion to the extended family, the nation, and the Emperor as its personification. But it should be noted that this is merely a highly refined form of the doctrine of consecration and self-control (*bushido*) which in other forms had dominated the cultural, military, and political life of the Japanese.

Throughout this seven-league-boots procession that we have just made through the psychologies of India, China, and Japan, is the problem of individuality, certainly a primary problem for the Western world, fundamental in Western European and American thought for the last two hundred years, and central in a thousand social and political discussions of the relation of the individual's life to the fulfillments and sacrifices that are involved in group living. It may be that the mass and class problems of the Asian civilizations had relatively little place for the psychology of individuality because of the essential structure of family, clan, class, and state. Perhaps these issues will become more clear and concrete as one works through, at greater depth, the material selected above from the psychologies of India, China, and Japan.

If one allows oneself to characterize in broad strokes the major differences between Asian psychology and European psychology, it is probable that the Asian approaches will seem "pessimistic" as far as the realization of a sound mind and a good life is concerned. At least, in striking contrast to the Greek belief in the vitality, beauty, and rightness of body and soul, and despite much asceticism and unworldliness in both traditions, the West offers a gusty and lusty belief in the possibility of realizing, through wisdom or common sense, some modicum of happiness here below. The Asian psychological approach is rather consistently skeptical of this. Whether we take the disparagement of the body and of the world as we find it in the Upanishads, or the retreat into an inner discipline as expressed in the Bhagavad-Gita and in yoga, or the belief that a middle course, involving Buddha's noble eightfold way, must lead one into a loss of individuality, a state of nirvana, we seem to find very little genuine faith that life is really good. We must proceed now to make some exceptions and clarifications.

First of all, we are dealing with philosophers, not with middle-level mentalities or with common people. It is the priestly caste and the warrior-caste-turned-philosopher that speak to us from ancient India. It is the disillusioned prince, Gautama, who speaks to the masses. The masses are actually reached in this way, masses comprising both men of princely birth and bearing and men of the humblest walks of life. The message of despair, as far as the here-and-now is concerned, is rather consistently brought out. One may indeed rightly object, as Professor Hsu has plainly shown (pages 145 ff.), indicating that most psychology and ethics in China were concerned with a practical, common-sense achievement of a way of growing, learning, thinking, and living that would make this life good. Moreover, we shall have to admit that the distinction we are trying to make holds for the *psychologies* of Europe and Asia. We are talking about optimistic and pessimistic *psychologies*; we cannot really assess the whole temper of whole peoples. But, up to a point, the receptiveness of China and Japan to Buddhism has its own symptomatic meaning. Despite the cheery mood of much Chinese literature and proverbial good sense, there can surely be no doubt that Buddhism made its way in China through the sort of otherworldliness that we have described, joining forces with the otherworldliness of Taoism. Perhaps a stronger case can be made for basic faith in life on the part of the more psychologically minded of the Buddhists in Japan, especially those of the Zen sect.

And here, we think, we find ourselves facing a deeper issue than the sheer question of the optimism or pessimism of European and Asian psychologies. We face the question whether there is actually a system of psychology in Asia which has a place for *human fulfillment in this world*. Stated in this way, it seems quite clear that there is very little place for this approach in the Asian systems. One can indeed refer to the bliss achieved by yoga and by Zen, but as we have already noted above (pages 98 ff.), there is a certain "double-talk" about these references to bliss, since it is the annihilation of affective states which is said to bring bliss, and one wonders if some double-talk is not fundamental to this whole psychological thesis. Indeed, there is no systematic cultivation of a life of happiness. The aims in psychological discipline are negative; that is, they are directed to getting rid of the overwhelming misery, suffering, and frustration which is conceived to lie in the very nature of human existence.

By virtue of this sharp contrast between the ocean of distress and the tiny droplet of serenity or bliss, one might very well refer to the enormous importance attached, in the Asian world, to self-control, to discipline, and to the limitation of immediate gratifications. One would expect—and indeed one finds—an apotheosis of self-control disciplines.

This prompts us to look more closely and see whether self-control is not actually more than a purely negative form of self-realization. The language of the mystics, whether in Hinduism, Buddhism, or Taoism, suggests that there was widespread realization, among disciplined adepts, of states of real joy or bliss. These experiences seem to have involved, on the one hand, enormous emotional investment in the world as a whole, a state of sublime identification with the cosmos of the sort called by Richard Bucke *cosmic consciousness*; or states of deep abrogation of individual competitiveness in favor of a loss of selfhood. Cosmic consciousness and loss of selfhood may, in some individuals, be identical, and in other individuals be different aspects of a general change in "state of consciousness," involving a reduction of striving and conflict.

So perhaps we should say that the Asian psychologies throughout their history have sought two chief positive goals: (1) a sense of the goodness of the ultimate timeless and limitless universe; and (2) on the other hand a sense of the unlimited ecstasy that could be found within a self which is freed of personal strivings. Again, we have stated these two doctrines in such a way as to suggest that they tend to coalesce; indeed, in most of Asian psychology they are, in fact, one: The goal is freedom from individual frustration and suffering, a process possible only when the battle of self against non-self is given up and cosmic order reigns within as it does without.

This is not a logical, orderly, scientific doctrine about human nature. It is a way of looking at human life which has developed among countless thinkers over great eons of time, over much of the earth's surface. It would be of interest and importance to the West even if it were limited to Asia; but it is not limited to Asia. In other volumes to follow, we shall attempt to show how quite different goals and means toward these goals have been developed by men of other cultures and other ways of understanding.

Another broad difference between Oriental and Occidental psychology seems to us to lie in method; specifically, in the way to seek psychological truth. One may either *seek diligently* and in travail the nature of psychological reality, or one may simply keep one's eyes open for it and *let it come* as it will. There are exceptions to be noted, but it will be apparent that, in the Upanishads and in yoga, a long and arduous *discipline* has to be pursued. The discipline of Buddha, though simple and of a common-sense variety, led on quickly throughout the East to the cultivation of a higher discipline having much in common with the Hindu methods. But whether we study it in Southern or Southeastern Asia, or in China or Japan, such discipline involves enormous feats of prolonged self-control. This notion of discipline appears particularly clearly in the interaction of systems within China, where, for example, the sage Confucius looks everywhere, under every stone, behind every star, so to speak, for "the Way," and Lao-tzu gently rebukes him for not finding it (pages 155 ff. above); and it appears in ancient and modern Zen in Japan in the form of a cultivation of that special state of mind, *satori*, in which one can *really* observe reality and self. Psychology that is worth anything is conceived to be very hard work.

But that is just one approach to psychology. It is, of course, true that the early Greek philosophers and the Church Fathers and Renaissance philosophers who followed them likewise had their discipline. In the marketplace, however, the Greeks *played with ideas*; they toyed with observations of self and the world. Whimsy and paradox played a large part in their thinking. Aristophanes was as much of their spirit as was Euripides. Often, as the Greeks said, "the god" breathes the ideas into one's understanding; just as, later on, the Christian deity, or an archangel serving Him, brought a message in a dream, and "sudden inspiration" became the hallmark of genius.

One of the differences between the sudden inspiration of Buddha and the sudden inspiration of Archimedes lies at least partly in the fact that Buddha had been seeking *within*, and seeking very hard indeed, while Archimedes was seeking *without*. It might, at first sight, appear that Oriental psychology is, from the beginning, predestined to look for *inner* realities; and Western psychology for *outer* realities. This will not, however, wholly solve our problem. As we have already seen (page 145), the otherworldliness or spirituality or mysticism of India is not truly parallel to any such trend in China or Japan. Chinese and Japanese life is rich in scientific and technical achievement, and Western psychology through much of its history is as introspective, as inclined to turn inward, as is the Oriental. No, the real distinction does not lie in inwardness or in outwardness, but in the conception of orderly discipline and training, guiding one into the recesses where true psychological reality can be directly observed—as contrasted with the rather matter-of-fact way of the West, well phrased by Thomas Hobbes when he asked the reader, when doubting what the author has to say, to "consider, if he also find not the same in himself." It would not take a British reader ten years of discipline to find out whether his own internal world was like that of Hobbes. When we teach Hobbes, our students accept or reject him on the first confrontation.

No, the Oriental psychologist carries his laboratory within him. It is the world of the spirit which he believes he can slowly learn to observe more and more accurately. He may, "sitting under the bo tree," have a great revelation; but in Buddha's case it came after years of strenuous seeking; and the first thing he did among his followers was to explain in detail and set up for them a discipline which, over the years and centuries, became more and more exacting, orderly, systematic. There was, in the foothills of Bihar, in northeastern India, a huge Buddhist university in which all knowledge was systematically portrayed, as it was shortly thereafter in the great monastic schools of Cluny and the other medieval academic centers of the West. But the force of the Buddhist tradition lies in *very arduous self-examination,* whereas the force of the Western educational tradition lay in training in *direct observation* and in the authority handed on from the past which taught one where to look. It did not lie in the training of mind and heart, as it did in the East; rather, it lay in training in the acceptance of what appeared to direct observation. The one thing of which Hinduism was afraid—direct outer observation, as full of illusion—was the thing in which Western intellectual history has excelled. So, after three centuries of observation of the outer world in a form which we know as science, we are coming around to a psychology which uses knowledge of the external world—physics and physiology—as a primary clue to the methods most suitable for the study of the world within.

Thus, paradoxically, we come through physiological relaxation or through the chemist's gift of many new drugs to the cultivation of those extraordinary states of altered awareness, cosmic consciousness, or depersonalization, which had been known, and indeed directly cultivated, for eons of time in the East. Instead of twenty years of discipline, it may take only a few hours or days of preparation—physiological and psychological—to prepare a person for the "unutterable revelations" which come from specially induced mystical states. The basic human nature which is tapped, displayed, or distorted, or whatever you believe happens during such conditions, is apparently much the same in the East and in the West, but the East has made these special states, so far from immediate, everyday, palpable reality, something to be *sought;* the West hands them to the aspirant or the casual inquirer as a gift from biological chemistry.

This is not to prejudge what kinds of psychology can best be learned in what parts of the world, and by what methods. "East is East, and West is West," and the two are coming together very fast indeed.

231

# Index

232

# Index

assuage (v) - to lessen (pain, etc); allay; calm; to satisfy
modicum (n) - some small amount, bit